UK COMMUNICATION STRATEGIES
FOR AFGHANISTAN, 2001–2014

For Claire

UK Communication Strategies for Afghanistan, 2001–2014

THOMAS W. CAWKWELL

Routledge
Taylor & Francis Group

LONDON AND NEW YORK

First published in paperback 2024

First published 2015 by Ashgate Publishing

Published 2016 by Routledge
4 Park Square, Milton Park, Abingdon, Oxon OX14 4RN

and by Routledge
605 Third Avenue, New York, NY 10158

Routledge is an imprint of the Taylor & Francis Group, an informa business

Publisher's Note
The publisher has gone to great lengths to ensure the quality of this reprint but points out that some imperfections in the original copies may be apparent.

British Library Cataloguing in Publication Data
A catalogue record for this book is available from the British Library

The Library of Congress Catologing-in-Publication Data has been applied for

ISBN 13: 978-1-4724-7352-3 (hbk)
ISBN 13: 978-1-03-292770-1 (pbk)
ISBN 13: 978-1-315-54944-6 (ebk)

DOI: 10.4324/9781315549446

Contents

Preface   *vii*

1   Strategy, Communication, and 'Strategic Communication'   1

2   Transnationalisation, Transnational Dilemmas, and
    Strategic Communication   13

3   The Rise of the Stabilisation Narrative   37

4   The Fall of the Stabilisation Narrative   59

5   The Counter-narcotics Narrative   87

6   The Counter-terrorism Narrative   115

7   Conclusion: The State of British Strategy and the
    Utility of Strategic Communication   149

Bibliography   *171*
Index   *195*

Preface

The difficulties faced by the United Kingdom in realising its stabilisation objectives in the War in Afghanistan (2001–2014) have precipitated a change in rhetorical approach by successive British Governments, from one based on liberal normative principles to one that emphasises traditional, rationalist precepts of 'national security interests'. This transformation of 'narrative' is identified in this work as chronologically analogous with the institutionalisation of 'strategic communication' practices and doctrine emanating from the defence establishment of the British state. In this work, I argue that changes in narrative approach and the emergence of strategic communication can be understood as a consequence of an overburdened British state attempting to free itself from a 'transnational dilemma' (King 2010): that is, to find a means of appealing coherently and succinctly to the benefits of participation in collective security whilst avoiding threatening the viability of collective security membership by acknowledging its costs. This transnational dilemma has been exacerbated by intra-state competition over the material and ideational aspects of British strategy in Helmand, and is traceable by close empirical analysis of three competing 'policy narratives' for Afghanistan: stabilisation, counter-narcotics, and counter-terrorism, respectively. Intra-state competition can, in turn, be conceptualised as the result of embedded inter-state relationships of political obligation and military cooperation referred to by Edmunds (2010) as the 'transnationalisation' of defence policy. UK policy in Afghanistan has been guided by transnational issues, specifically the maintenance of NATO as a collective security apparatus and of the 'special relationship' with the United States, through which Britain secures and projects its national interest. I argue that the UK's grand strategic commitment to transnationalisation underscores an 'unstatable' ultimate policy of meeting the expectations of the United States and NATO, and that the development of various policies and narratives for Afghanistan can be understood primarily in such terms. In Afghanistan, transnationalisation and the concordant pursuit of satisfying American and NATO expectations has come at the cost of a significant divestment of strategic autonomy, which has uprooted traditional, nationally-based concepts of strategy and policy to the transnational level and resulted in a strategic vacuum wherein intra-state competition has flourished. This, I argue, has compromised the ability for Britain to link policy to operations (to 'do' strategy) in Afghanistan, a point which can be empirically measured by reference to the discordant and contradictory aspects of aforementioned policy narratives, which have been rooted in the institutional interests of various elements of the state. Strategic communication has arisen out of this situation as a means for the state to overcome the transnational dilemma by promoting a unified 'strategic narrative' for Afghanistan that has reconfigured the narrative for the conflict to one that emphasises the conflict not in terms of collective security but in 'national' terms. This work concludes by arguing that, in sidestepping rather than confronting the core dilemmas of British strategy, the emergence of strategic communication can be seen as posing as many problems as solutions for the UK state.

Chapter 1

Strategy, Communication, and 'Strategic Communication'

In the last weeks of 2014, Britain's war in Afghanistan came to an end. Since the terrorist attacks of September 11, 2001, the world witnessed 13 years of intervention, making the conflict one of the longest in the modern histories of all of its participants. Alongside the United States and the United Kingdom, the conflict involved a coalition of 48 other states within and without the UN-mandated International Security Assistance Force (ISAF), which in turn ran alongside the US-led Operation Enduring Freedom (OEF) tasked with decimating or destroying international terrorist networks and stabilising the country. As the conflict wore on, strategies and operational approaches moved from narrow counter-terrorist missions to 'population-centric' counterinsurgency operations, and spread beyond the frontiers of Afghanistan into neighbouring Pakistan. The war proved costly for many of these contributors, with over 3,000 coalition deaths between them. For the UK, Afghanistan has resulted in 454 troop deaths and over 2,000 wounded in action and, according to one estimate by the Royal United Services Institute, came at a material cost of around £20 billion (Wright 2014, online). Statistics are scarce regarding the number of dead insurgents, but is thought to be in the tens of thousands (Dawi 2014, online). Similar figures must be considered likely in relation to Afghan and Pakistani civilians and security forces.

Given the magnitude of the intervention and the depth of British involvement, it is natural that questions about the efficacy of the mission became commonplace as the war concluded. What effect did intervention have, and to what extent was it successful? Of course, any answer to such questions must be prefaced by the simple fact that history is still being written on the fate of the country and its state, and whether the state will be able to maintain or expand its security writ over the country rests largely on its ability to consolidate its own authority and reach settlements with competing power bases in the country. Additionally, any answer regarding what was achieved in Afghanistan depends almost entirely upon how one understands the purpose of the mission and, from that, how one measures 'success'. In the waning years of the conflict, political leaders of states involved in Afghanistan employed 'narratives' that claimed that the mission was successful, and did so by emphasising the baseline successes of preventing Afghan-centric international terrorism and the building up of the Afghan National Security Forces (ANSF) as a means of continuing such efforts in anticipation of ISAF's departure at the end of 2014. Indeed, by this preferred metric, one may observe considerable achievements. At a minimal level, intervention in

Afghanistan realised its immediate goals: first and foremost, al-Qaeda have been largely dismantled or at least pushed out of Afghanistan, and the ability of Islamist extremist groups to plan and/or carry out terrorist attacks from the country have been largely mitigated for the time being. Secondly, the Taliban were removed from power and, whatever the threat they pose to the current Afghan state, it is unlikely that they will ever recover (as currently constituted) the power they held over large parts of Afghanistan as they did prior to September 2001. This is so in no small part due to a third achievement of fostering the improvement of the capabilities of the ANSF, which ISAF trained and which took over responsibility for Afghan security in time for ISAF's drawdown.

These accomplishments are somewhat muted, however, when one considers that the first two were achieved within months of the beginning of the intervention in October 2001. Indeed, the majority of the history of the international community's involvement in the war in Afghanistan is a story of efforts not directly related to counter-terrorism. Rather, ISAF was primarily concerned with 'stabilising' the Afghan state and assisting it in securing and buttressing its authority over its territory. To present the 'end-game' of Afghanistan in purely security terms, as is now the wont of statesmen and women across the coalition, is to provide a highly revised and reductive account of the campaign but, crucially, also an instrumental one that to some extent justifies the costs of intervention. The problem with such narratives is that, in order to portray the Afghan mission as ultimately successful, they omit large parts of the history of the conflict. ISAF's mission was far broader than the narrow national security retrospectives many officials now focus upon, so much so that the stabilisation mission in Afghanistan expanded over time to include commitments to practically every conceivable area of state jurisdiction, from the sustainability of its security forces to the development of a myriad of governance capabilities. Such efforts have, in effect, constituted ISAF's strategy for Afghanistan, and on this basis there are significant grounds to question the claims of success made by Government officials.

Despite (or, as I will argue, reflective of) the revisionist narratives of today, delivering stabilising effects has been a fundamental strategic problem for the UK over the last decade, and is one that calls into question the viability of the operating model. Ultimately, Britain and ISAF's strategies for Afghanistan were dependent on the ability of the Afghan state to resolve its own internal problems, leading to a general strategic dilemma regarding the suitability of stabilisation, specifically of relating the use of military force to achieve what was, in essence, a political objective. At the same time, there appear few options available to states in their efforts to prevent terrorist attacks other than stabilisation (Paris 2010: 340). However one assesses the merit of ISAF's presence in Afghanistan, it should be clear that the strategic methodology by which it sought to facilitate the stabilisation of Afghanistan held significant flaws. For the UK in particular, experiences of Afghan stabilisation have precipitated a crisis of confidence in its ability to devise strategy to meet the demands of an increasingly unpredictable world. The difficulties Britain and its partners have faced in stabilising Afghanistan have manifested themselves in discursive terms – what state officials say – in a way that downplays much of the norm-based content of the stabilisation agenda. This work is about understanding this revisionism as a consequence of the difficulties of strategy in the contemporary era; specifically, how communication has been shaped by strategic dilemmas.

This introductory chapter provides context to this issue by inquiring how these problems of strategy came about. It argues that the root problem for the UK lies in the institutional and normative conditions in which stabilisation and state-building models have been inculcated; specifically, a divestment of strategy above the national level and a limiting of strategic possibilities to liberal normative options because of the way collective security mechanisms work. This has led to a conceptual blurring of ideas of interests and values and of 'collective' and 'national' which, I argue, has fundamentally affected British strategy in Afghanistan by confounding the terms of reference by which strategists do their work. A traditional, linear understanding of strategy begins at the identity and interests of an actor or set of actors and proceeds from this starting point to devise policies, from which strategies, operations and tactics are made. Under circumstances where the actors or interests are not easily divisible, strategy can become confused and unclear and, what is more, cannot be easily articulated. It is in this context that so-called 'strategic communication' (SC) institutional processes and practices have emerged as a discursive means of 'plugging the gap' in strategic thinking. It is this instrumental use of communications that makes it ostensibly strategic. In the chapters that follow, I will argue that SC can be understood as a response by the UK state to the problems of collective security strategy in Afghanistan typified by 'transnational dilemmas'. I then apply this concept to the various policies pursued by the UK in Afghanistan.

The State of British Strategy

The UK has in recent years undergone a period of self-reflection and recrimination concerning its apparent lack of strategic capability. Criticism of Britain's strategic acumen coincided with an expansion of UK activities in Afghanistan since 2006 and the emergence of subsequent operational shortcomings. Indeed, one may quite reasonably deduce that, given the centrality of Afghanistan to British defence interests over the last decade and a half, it is the war in Afghanistan that has served as the crucible for the British state's realisation of its strategic limitations. This view is borne out by reference to chronological comparison of Britain's struggles in Afghanistan with the prevalence of critical reports from within the UK's political establishment regarding Government's approach to the conflict, as well as with scholarly articles in recent years that have questioned the viability of the UK's strategic posture (Strachan 2005, 2006, 2008). In February 2007, Field Marshal Sir Peter Inge voiced his concern that Britain had 'lost the ability to think strategically' (Betz and Cormack 2009: 320). This sentiment was echoed by then-

Chief of the Defence Staff Jock Stirrup in December 2009, when he declared that Britain had 'lost an institutionalised capacity for, and culture of, strategic thought' and offered the advice that '[a]ll we do at the tactical and operational level needs to be rooted in good strategic soil, and therefore in our national interest' (Stirrup 2009, online). In October 2010, the House of Commons Public Administration Select Committee released its review of Britain's strategic capabilities, titled 'Who Does UK Strategy?', which opined that Britain has 'simply fallen out of the habit' of all things strategic (PASC 2010: 3). In that report, the Committee drew upon the testimony of Sir Robert Fry, who argued that Britain had

> [fallen] out of the habit of Grand Strategy, and I think that is what happened to us in the second part of the 20th century. Also larger strategies that were extra-national—so NATO, the cold war – took over and really took the place of any Grand Strategy. (2010: 14)

Both Stirrup and Fry's remarks contained within them an intimation that Britain's difficulties with strategic thought were – at least in part – the result of the interests supporting them not being sufficiently 'national', seemingly as a consequence of a divestment of strategic responsibility from the national to the transnational level, and, in Fry's case in particular, a concomitant hardwiring of collective security logic over and above any semblance of a distinctively 'national' strategy. Furthermore, for Fry the atrophying of British strategic autonomy is nothing new; rather, it has been a multi-generational process spanning at least the last half century and most of the history of NATO. This point raises an important question, however: if Britain has been operating under such logic for so long, why is it that this alleged divestment of strategic authority to the transnational level has only recently produced an awareness of a fundamental 'loss of capacity' for strategy? The institutional makeup of British defence policy is clearly important in deducing 'who makes UK strategy', but it does not provide one with an answer as to what animates the strategic process at the level of organising concepts.

Providing such an answer therefore requires an analysis of what is conceptually different about contemporary strategic circumstances in relation to those of the Cold War, which is of course the emphasis placed on combatting non-state terrorism as an international security issue of the highest order. Along these lines, the most enduring strategic dilemma for the UK in the twenty-first century has been determining the best response to international terrorism. For illustrative purposes, it is worth considering the UK's responses to this question in terms of a rough binary between liberal and illiberal approaches. The first approach, in line with liberal peace theory, is to extrapolate that failed states breed terrorism, and that the problem of terrorism is foundationally the result of political and economic illiberalism (Doyle 1986; O'Neal et al. 1996). From this premise, it becomes incumbent to posit that the long-term solution to terrorism is political and economic reform (generally in the guise of liberalisation and democratisation), and therefore to develop policies and strategies that focus on underlying issues –

that is, to treat terrorism as a symptom of a broader diagnosis of socio-economic illness stemming from weak governance and underdevelopment. The second approach to global security is combine comparatively straightforward counter-terrorist activity, typically consisting of air strikes and limited covert operations in conjunction with the development and maintenance of indigenous security apparatuses and forms of state governance (West 2011; Gentile 2011a, 2011b; King 2010a; McCrisken 2011). These positions typically give greater credence to the utility of military force than those of the stabilisation and human security schools. Articulations of this kind have grown in frequency amongst commentators and state officials in recent years, and similar sentiments can be observed in the revisionist explanations of politicians who increasingly framed the Afghan campaign in sparse counter-terrorist language. Indeed, this book will make the revised character of this approach quite clear, since the emergence of a more limited counter-terrorist retrospective for Afghanistan can be traced as taking place in conjunction with the gradual disappearance of liberal normative explanations.

Understanding these two perspectives in binary terms is useful because British strategy for Afghanistan (and the way in which strategy has been communicated) gradually but (so I will argue) decisively moved from the first approach to the second between 2001 and 2014. We may see this movement as an attempt by the British state to simultaneously extricate itself from less successful aspects of liberal state-building in Afghanistan and to maintain – and possibly strengthen – its domestic argument for staying in Afghanistan and, by consequence, staying true to its obligations to the collective security framework of NATO. The story of Afghanistan, simply put, is that British strategy – grounded as it is in collective security frameworks and bounded by liberal normative precepts – became confounded by the state's attempts to meet the challenges of international terrorism by over-rationalisation of the means (stabilisation) by which to do so. This was so to the extent that the stabilisation of Afghanistan at points appeared to trump counter-terrorism aims, and did so because the collective security mechanism of NATO operates on the basis of a common denominator of liberal institutionalism. Stabilisation was, in this sense, strategically appealing to NATO members because it reflected their own assumptions about how their individual and collective power should be wielded. As both a leader and follower of institutional and normative principles, the UK has for much of the last two decades locked itself into a line of thinking that declares that the best way to respond to the security challenges states face in the contemporary era is through concerted action that places its faith in liberalisation and democratisation. More specifically, Britain relies on collective security actions coordinated around and constituted by liberal frameworks. What is significant about Britain's defence posture is that it implies a divestment of autonomy to the transnational level and, therefore, a significant narrowing of strategic options that has extended only to those that accord with liberal principles.

Thus, the problematic of strategy expressed by Fry, Stirrup and others within the British establishment – that strategy must be grounded in national interests – is one made more complicated because British strategy appears to be driven by a mixture of collective security interests and liberal interventionist principles. The result is that terms like 'national' and 'interests' have lost much of their utility as foundational nomenclatural units of strategic theorising, as Britain's adherence to strategic thinking within the purviews of collective mechanisms and liberal principles has progressively eroded ideas of the 'national' and 'interest' over the last two decades, to the point where they are understood by policymakers as indistinguishable from the collective security mechanisms and liberal principles through which British strategy is made. As such, a national interest is also a collective interest, and vice versa; what is a value is an interest, and vice versa; and so on. The terminology of strategy has lost its distinctiveness, and so thinking about strategy has become more challenging; and this has occurred because of the socialising aspects of collective security membership, which provides the UK with the means by which – and apparently only by which – its 'national' interests can be realised. Because British national interests – tied up as they are in the promotion of the rule of law and liberal norms – are difficult to separate from those of the international community, and because to set about separating those interests would be in many ways antithetical to Britain's interests, they can be said to have become 'transnationalised' (Edmunds 2010). As such, it would appear that the options for reconstituting Britain's interests as *nationally-based* are practically and conceptually limited. The blurring of values with interests and 'national' with 'international' indicates that Britain's 'interests are not defined well enough to impart a meaningful idea of strategy' (Ritchie 2014: 88). This issue of definition is key: if one cannot easily define where 'national' interests end and 'transnational' interests begin, how can one hope to devise a distinctively 'national' strategy? If strategy exists beyond the level of the state, improving strategic capabilities at the level of the state will be a difficult task. Again, the issues informing British strategy are complex and multi-faceted. This work does not aim, nor is it designed, to provide an account for or a solution to all of these issues. Rather, it focuses on why strategy is so difficult (and difficult to articulate) for the UK and how strategy for Afghanistan, devised under conditions of collective security, has been communicated. More precisely, this work is interested in the historical development and performance of 'strategic communication' practices by the British state in relation to the war in Afghanistan and how these practices can be understood as consequences of the shaping effects of collective security on British defence policy.

Thesis and Structure

This work views the role of Afghan and ISAF/NATO politics as crucial in determining the British approach to explaining its strategy for the Afghan conflict. Britain's difficulties in explaining its purpose in Afghanistan is intimately wound up in its role there within a collective security apparatus. As such, if communication about strategy is naturally dependent upon the substance of the strategy itself (its 'explainability'), it is worth considering the difficulty in communicating the purpose of British involvement in Afghanistan as a consequence of a cognitive shift away from a traditional continuum of state activity of interests, policies, and strategies and toward a multi-national or inter-state continuum, wherein one may locate a significant divestment and blurring of 'national interests' to the transnational level. This in turn may be understood as being a result of an adherence to the internationalist assumptions implicit within the liberal norms that bind Western collective security mechanisms together. Anthony King has described this communicative difficulty in conceptual terms as a 'transnational dilemma' wherein mid-level states such as the United Kingdom have found it difficult to articulate to the British public the purpose of collective security missions like that in Afghanistan in collective security terms (2010b: 388). King encapsulates the dilemma thusly:

> States are now increasingly interdependent and can increasingly defend themselves only by cooperating with other polities; individual security and defence is dependent upon the generation of collective security. Consequently, since the end of the Cold War, states have been driven to contribute to military expeditions which do not seem to be in the direct national interest but from which they cannot exempt themselves for fear of being excluded from access to critical shared security goods. (2010b: 388)

Directness of interest is pivotal in understanding the transnational dilemma. Collective security arrangements naturally militate against 'direct' effects, since interests and threats are (ideally) joint and several between all members. Indeed, this principle is enshrined in Article V of the NATO Charter, a point referenced by NATO in response to the 9/11 attacks. There is, therefore, an institutionally built-in predisposition within NATO to accepting the principle of indirectness; in a very real sense, it is its defining precept. Explaining collective security accurately requires some account of the inherent 'indirectness' of such frameworks, but to do so risks a level of abstraction perhaps unamenable to everyday political discourse. This is the *quid pro quo* of alliance membership described by King. The dilemma is a communicative one at root, and is one that confounds easy explanation. What Britain's rhetorical response to the hardships of Afghanistan demonstrates is that the UK has largely sought to avoid this dilemma in both its liberal institutionalist (stabilisation-centric) and realist (counter-terrorist) articulations of its Afghan policy. In the first instance, it sought to claim that national interests were subsumed within and inseparable from collective interests. In the second instance, it argued that collective interests were subsumed within – but subordinate to – the national interest. Although, as we shall see, these two approaches are in many ways incompatible with one another, they do share a common feature of seeking to avoid at all cost the notion that collective and national interests can at times be quite separate. Even though a mutual and reciprocal acceptance of the principle of indirectness of interest is quite obviously a key element in the maintenance of any collective mechanism it is, in times of military hardship, paradoxically also a truth that apparently (for the UK, at least) must never be spoken, lest the tenuous nature of inter-state diplomacy be revealed and the strategic calculus of collective action (that is, the sharing of costs and benefits) therein exposed to question.

Avoidance of articulating the difference between two sets of interest – 'national' and 'collective' – should be understood more generally as an aversion on the part of Governments to revealing the relatively frail ties that bind states together within collective security mechanisms to the prevarications of public opinion. This is so particularly during periods where the activities pursued by such mechanisms do not appear to be providing sufficient collective goods to its constituent members, or where those goods appear to be outweighed or otherwise mitigated by the associated costs. Such aversion suggests that states are (a) aware of the importance of public opinion and the need to shield certain diplomatic arrangements from the dangers of debate in the public weal, and (b) that some diplomatic arrangements – particularly those that cut to the heart of supposed 'national' interests (in this case, the maintenance of collective security apparatuses) are possibly considered too important to be subjected to such debate. A view expressed in this work is that Afghanistan has tested this arrangement because it has been so difficult a campaign. This, incidentally, is why Afghanistan is also the ideal case study of investigating national strategy under collective security: it was the quintessential liberal institutionalist conflict of this century.

It was within this context that 'strategic communication' (SC) doctrine arose within the UK defence establishment as a potential corrective to the communication excesses wrought by the state's avoidance of the transnational dilemma. SC as a process came into being for the purposes of streamlining dispersed, institutionally-aligned messaging on Afghanistan within a more concentrated rubric of 'nation-centric' defence interests, thereby reducing the potential for mixed messaging (referred to by strategic communicators as 'information fratricide') and facilitating the establishment of a single, coherent narrative for the conflict. At the conceptual level, SC is dubbed 'strategic' because it is held to possess an instrumental quality to it, a point borne out by the definition provided by the Ministry of Defence (MOD) in its *Joint Doctrine Note 1.12*, which states that SC's utility is in '[a]dvancing national interests by using all means of Defence communication to influence the attitudes and behaviours of people' (DCDC 2012: 3-1). In this specific context, SC processes and practices have been largely successful; as this work will show, instances of defining the purpose of British involvement in Afghanistan as anything other than a vital 'national security interest' concerned with preventing acts of terrorism in the United Kingdom drastically diminished following the introduction of SC in late 2008 and early 2009. Public support for the Afghan mission actually increased after 2009, possibly as a result of more focused messaging (Kriner and Wilson 2010). Most importantly, through implementation of SC doctrine, UK communication efforts can truly be seen as 'strategic' in the respect that it has seemingly 'solved' the transnational dilemma: state communication is now reflexive, or self-consciously aware, of the confusing nature of 'indirect threats', the institutional ambiguity within state bureaucracies that indirectness can foster, and the negative effects of mixed messaging, and as such its communication efforts are now oriented around preventing such confusion by focusing on greater coordination of what can be said and on avoiding what cannot – i.e., that which emphasises the indirect and contingent nature of collective security.

The perspective offered in this book represents a departure from the growing literature on strategic communication. Typically, these works are prescriptive in nature – they focus on the potential utility of SC and assume it to offer solutions to some of the strategic issues states like Britain face (Helmus, Paul and Glenn 2007, Tatham 2008, Mackay and Tatham 2009, Cornish, Lindley-French and Yorke 2011, DCDC 2012). They tend to assume that SC is capable of adding significant value to understandings of contemporary strategy because they claim that much of the landscape of war is now discursively oriented. Much of this literature bases its assumptions on the Rupert Smith's assertion of the diminishing 'utility of force' and the increasing importance of communication *vis-à-vis* physical, or 'kinetic' effects in winning 'hearts and minds' – both domestic and in-theatre – to achieving success in stabilisation operations (Smith 2006). In a few specific cases, SC and 'narratives' more generally have been presented along such lines as not only a solution to the difficulties of traversing the transnational dilemma, but also as having the potential to re-define the nature of counterinsurgency operations and effectively lead strategy (Roenfeldt 2011, Simpson 2012). This work is critical of such assumptions. Whatever the merits of their approaches, it is my contention that they arrive at such conclusions because they do not bother to problematise SC. They tend to focus on theoretical principles of SC or empirical case studies of its effects at the operational level of war and extrapolate the strategic significance of communication therefrom. They do not adequately analyse SC as a specific, empirically isolatable historical and social phenomenon, and as a result fail to make the connection between the problematic of British strategy and SC as a consequence of the communication issues associated with the transnational dilemma. Indeed, despite a great deal of scholarship on the state of British strategy and the utility of SC, there has thus far been little work that analyses the linkages between them.

This work seeks to make headway in this respect by conceiving of SC as both a consequence of and response to communication issues arising from the transnational dilemma and the transnationalisation of British defence policy more generally. This chapter has outlined the strategic issues faced by British policymakers and how prevailing institutional and normative orthodoxies contribute to the complexities of statecraft for mid-level powers like the UK. I have argued that the strategic challenge of the last 15 years has been that of terrorism and failed states, and have introduced the idea (to be further explored in the next chapter) that the institutional and normative responses to these challenges has produced a scenario where states belonging to the collective security apparatus of NATO – the UK being the object of study in this case – have been socialised within that system in a way that has caused them to internalise collective security institutional interests and norms, a phenomena best described as the transnationalisation of defence policy (Edmunds 2010). This has in turn facilitated the conditions for transnational dilemmas to emerge and mixed messaging to ensue, often as a consequence of intra-state rivalries caused by a lack of autonomy in policy and precipitated by a concordant loss of strategic direction. States within ISAF and state departments and institutions within the UK have tended to pursue their own organisational aims; the latter as a means of contributing to Britain's 'comprehensive approach' to fulfilling its transnational obligations in Afghanistan. The result for the UK has been the emergence of several policies or strategies – counter-terrorism, counter-narcotics, and stabilisation, namely – that reflected the institutional dynamics and interests from which they were created. As both a consequence of and response to this scenario, SC doctrine and processes can be seen as an object of inquiry that illuminates how the British state has responded in narrative terms to contemporary strategic challenges. The transnational dilemma as defined by King should be understood as that which makes strategy difficult to articulate; SC, in its attempt to simplify the complexities of strategy, can therefore be seen as a response to the transnational dilemma. This work is, therefore, animated by the following questions: *how has the transnational dilemma affected British strategy in Afghanistan, and how has it both necessitated and complicated SC efforts for the war in Afghanistan?*

In order to justify my answer to these questions, it is necessary to set out the parameters for research. In the main, this means determining the relationship between transnationalised defence policy and transnational dilemmas, and between transnational dilemmas and SC practices. Naturally, defining each of these terms is a prerequisite to this end. In this introductory chapter, I have provided definitions for and outlined the relationship between the latter two terms. In the next chapter of this work, I give similar treatment to Tim Edmunds' transnationalisation thesis and make a more detailed case for viewing Britain's defence policy and strategy in Afghanistan as being subject to processes of transnationalisation. I do so by reference to existing literature on the subject and by way of a brief comparative assessment of the UK's participation in the War in Afghanistan with that of the Second Afghan War of 1878–1881. I then relate the theoretical and empirical lessons from these analyses to a brief summary of the political effects of transnationalised policy on the UK state's institutional dynamics in Afghanistan, thereby providing an appropriate grounding for my analysis of empirical evidence in Chapters 3–6. These chapters detail the evolution of three distinct yet interconnected 'strategic narratives' – those of stabilisation, counter-narcotics, and counter-terrorism, respectively – employed by the UK to justify and explain its strategies in Afghanistan. These narratives demonstrate how British strategies – and the communication of those strategies – have struggled against both the direct and indirect effects of transnationalised defence policy, specifically the pressure placed on the UK by the United States to perform, and a concomitant tension within the British state to meet the requirements of such demands. These pressures have produced contradictions in strategy by causing the UK to pursue incompatible aims and, as a result, have exacerbated conflicts of interest between those tasked in delivering strategy. Such factors have exposed the difficulties of maintaining a consistent and coherent 'meta-narrative' for Afghanistan when its component parts (the three 'strategic narratives' of stabilisation, counter-narcotics and counter-terrorism) frequently failed to mutually support one another. This problematic precipitated the need for SC doctrine and process within the British state. The findings of the empirical chapters of this work support the idea of SC as a direct response to the transnational dilemma. In the concluding chapter of this work, I discuss the implications of connections between British SC and its collective security membership for the future of UK strategy.

Chapter 2

Transnationalisation, Transnational Dilemmas, and Strategic Communication

In the first chapter of this work, the central theme of 'transnationalisation' of defence policy was contextualised against literature on the UK's strategic capabilities and shortcomings. This chapter builds on the first by making the theoretical and empirical case for the 'transnationalisation' of British policy and strategy, which posits that the UK's capacity to formulate grand strategy is largely dependent upon issues of normative and material dependency centred on collective security frameworks. I argue in this chapter that Britain's national or grand strategy is to a great extent contingent and discretionary: it exists only in relation to other national strategies within the confines of an Alliance meta-strategy. This is a rather uncontroversial statement given that it is empirically defensible in terms of Britain's military track record, its numerous documents of strategic intent, and its commitment to transforming the structure and outlook of its Armed Forces in order to meet the demands of collective security cooperation. In operational terms, Britain has taken the leading role in supporting the three major American-led interventions of this century – Afghanistan, Iraq and Libya, respectively – and has participated in every such intervention since the end of the Cold War, with the exceptions of Somalia in the mid-1990s and the aborted intervention in Syria in 2013 (Dorman 2008: vii). In terms of politics and diplomacy, Britain orders its identity and interests in the world system around these inter-related pillars of collective security and the 'special relationship', epitomised in recent years by numerous strategic reviews and security strategies and, of course, in rhetorical terms by the former Prime Minister Tony Blair's commitment to stand 'shoulder to shoulder' with the United States (Blair 2001a, online). From a military and institutional perspective, these commitments have impelled the British Armed Forces to undergo a period of 'transformation' of its force structure and capabilities to one of 'expeditionary warfare' in recent years, both as an attempt to emulate and coordinate with similar undertakings in the United States (Farrell 2008; Edmunds 2010; King 2011). In these ways, it is possible to posit both the dependence of the UK on the United States for the maintenance of British defence interests and strategy and, more broadly, the significant shaping effects of transnational factors on the composition and behaviour of the British state.

This way of viewing the recent history of the United Kingdom's defence affairs is reflected in Edmunds' assertion that Britain is subject to a 'transnationalisation' of its strategic practices (2010: 378–82). For Edmunds, these external shaping factors are significantly responsible for the tensions and conflicts within the

British state as its various constituent parts (government departments and the military) come to terms with drastic institutional reform. In this chapter, I utilise and expand upon Edmunds' argument as a means of substantiating the transnationalisation thesis (as it applies to Britain's experiences in Afghanistan) and explaining the tensions and conflicts within the British state regarding the prosecution of that conflict. I situate Edmunds' work within a wider set of literature on military transformation and Britain's strategic dilemmas (Cornish and Dorman 2009a, 2009b, 2010, 2011, 2012; Farrell 2008; Farrell and Gordon 2009; King 2010a, 2010b, 2011a, 2011b). I argue that transnationalisation of strategic practice is a significant variable that has exacerbated intra-state tensions, as the various departments of state have repeatedly come into conflict over their respective remits and the overall direction and purpose of the Helmand campaign of 2006–2014. I contend that these institutional conflicts have contributed to the production of operational difficulties that exposed serious strategic shortcomings whereby each institution had, contrary to the demands of the 'Comprehensive Approach' (the UK's doctrine for inter-departmental coordination of effort), worked according to its own relatively narrow agenda at the expense of overall strategic clarity of purpose and operational cohesion.

Transnationalisation

In order to meet the demands of contemporary security, British defence policy depends on the collective mechanism of NATO and, more specifically, on its 'special relationship' with the United States. A cursory study of the core texts of British defence doctrine published in recent years plainly bears out this statement. The 2010 National Security Strategy, for example, states that the UK can 'best pursue [its] interests' 'through a commitment to collective security via a rules-based international system and our key alliances, notably with the United States of America' (Cabinet 2010: 10). 2011's *Joint Doctrine Publication 0-01: UK Defence Doctrine* states quite clearly that Britain is locked into collective security since the 'UK rarely can, or even should, act alone' (MOD 2011: 1–3). Similarly, the 2008 National Security Strategy, published under the Brown Government, states that the United States is 'central' to British national security and that collective mechanisms are 'the most effective way of managing and reducing the threats we face' (Cabinet 2008: 7, 8). The 2010 Strategic Defence and Security Review (SDSR) further elucidated this foundational aspect of British statecraft by stating in no uncertain terms that 'collective security through NATO' will be the 'basis for territorial defence of the UK' (2010: 13). Taken individually or collectively, the message is quite clear: the primary purpose for which the state exists – defence of the realm – is only fully realisable through collective security membership. This carries with it the implication that the imperative of optimising British power and influence in national policymaking by coupling it with other states requires a willingness to compromise a large degree of national autonomy to collective security mechanisms. Britain pursues its interests through a collective mechanism, and therefore its interests must be balanced with the interests of others. Bluntly put, the furtherance of UK interests is conjoined to those of the United States (and to a far lesser extent that of NATO): in placing collective security as vital and inalienable from British security, the UK has put itself in a position where it advocates, ostensibly for its own purposes, a 'transnational' policy agenda, one that articulates the common interests of all members. What the core texts noted above demonstrate is that collective action, central as it is to the realisation of Britain's 'national' interests, is not simply a statement of intent, but also a statement of identity. In meaningful ways, participation in collective security defines the UK's role in the world, for without the apparatus of collective action, its ability to act would be decidedly reduced. Belief in that core identity means that Britain must shape and project its own national interests within the milieu of collective security, requiring its full political and military participation in joint ventures such as that of ISAF in Afghanistan.

Alongside accepting the United States' role as ultimate guarantor of UK security, Britain's willingness to participate in American-led coalitions allows it to continue to 'punch above its weight' in world affairs, and goes some way in justifying its arguably disproportionate status as a permanent member of the United Nations Security Council. There are, therefore, substantial strategic advantages to Britain's transatlantic alignment, without which it is doubtful that it could continue to exert significant influence on the international system. The trade-off of this relationship, naturally, is that British strategy and policy is ultimately contingent upon American strategy and policy, limiting the UK's strategic horizons and rendering much of British foreign and defence policy discretionary and in all practical senses subservient to US interests (Freedman 2007; Gray 2008; Porter 2010b: 9). This dynamic is not *necessarily* problematic for the UK. While American and British national interests are not synonymous, they are, as Colin Gray has noted, 'close enough' (2008: 15). Both countries rely on the maintenance of the international political and economic order they largely created (through the creation of various laws, conventions, and international organisations of global governance) in order to best pursue their national interests (described by the 2010 National Security Strategy as consisting of 'freedom, prosperity and security' (2010: 10)). For its part, the United States depends to some degree on having strong diplomatic and military partners within the institutional architecture of NATO and the United Nations, whilst the UK is dependent (for the time being) on the United States for its nuclear deterrent and for much of its military and intelligence capabilities (Porter 2014: 130). Along these lines, a coincidence of British defence and foreign policy with that of the United States appears rather natural and fortuitous, even if from an analytical perspective this merger produces a transnational 'securitisation' effect of the two sets of interests becoming 'so interlinked that their [individual] security problems cannot reasonably be analysed or resolved apart from one another' (Buzan and Wæver 2003: 44). The benefits of the closeness of the US-UK special relationship come with obligations, of course; Britain has political obligations to the United States to participate in US-led operations, and this has produced a concomitant need for Britain to transform its armed forces to meet the demands of operational and tactical inter-operability with America's military (Betz and Cormack 2009: 324; Farrell 2008). This sense of obligation and the transformation process that both informs and supports it serve a specific purpose, so David Betz and Anthony Cormack have argued, of allowing the UK to 'show willing' to the United States by wielding its 'armed forces as an instrument of policy', i.e. as a means of maintaining the special relationship (2009: 333–4).

It is out of this milieu of transformation and obligation that Edmunds' identified the transnationalisation of strategic practice and defence policy. For Edmunds, Britain's commitment to securing its national interests by honouring its political obligations to the United States and transforming its military to that end evinces an iterative momentum, constituting further those commitments via processes of multinationalisation. What this means is that the more capable the military apparatus of the British state, the more its political obligations can be fulfilled. Likewise, the more numerous and onerous the political obligations on the state, the greater the need for transformation to meet those obligations and the greater the strain on the state institutions being transformed (Edmunds 2010: 382). Along these lines, Edmunds' analysis corresponds with that of Anthony King, who argues that the transnationalisation of national defence policies in Europe stems from the institutionalisation of guiding principles of interoperability within states' respective armed forces (King 2011: 10). This evolution appears to have received additional impetus in recent years as the demands of interoperability have increased as a consequence of economic downturn and the tightening of defence budgets (Cornish and Dorman 2012: 223). This cycle has been evident throughout the Afghan campaign, particularly in terms of the US repeatedly requesting British increases in troop contributions (and British acquiescence to those requests) to support American-led operations; thus, the dynamic between political obligation and military cooperation appears to be self-regulating or internally propelled. In cases where British civilian leadership appears to have been reticent about providing such support, as was the case with Gordon Brown in 2007, British military leaders responded with outrage that the maintenance of UK political obligations to Afghanistan might not be accompanied by sufficient military contributions to the collective security effort.

The obligation-cooperation cycle can be seen more generally in defence procurement. This, as Robin Porter has argued, is a material manifestation of Britain's desire for ever-improved coordination with the US: the UK buys from US suppliers to maintain its ability to remain interoperable with the US in operational terms. In doing so it fulfils its political obligations, and by participating in operations secures its place in the elite circle of claimants of collective security goods (2014: 120, 128). The same patterns apply in the embedding of intelligence structures between the UK and US: much of the former's intelligence gathering capabilities are now dependent upon US infrastructure (including that which is located within the UK's territories), lending itself to the perspective that Britain is obligated to appreciate US provisions and, more controversially, is subject to its policy positions being shaped by US influences in ways that 'may in fact be impeding the development of an independent British foreign policy' (Porter 2014: 130). In Porter's view, such institutional linkages quite clearly imply 'significant derogations of UK sovereignty' (2014: 130). This embedding further cements Britain within a transnationalised policy arrangement, since the tools through which policy is informed and actioned are increasingly collectively pooled and, crucially, not independent (either intellectually or operationally) from collective (specifically American) interests. By virtue of its iterative nature, the material relationships Britain has with other states produces further and repeated pressure on the state in terms of political obligation. The unfolding of this process explains why, according to Gray, a crucial precept of British defence policy and strategy is that it manages to 'satisfy US expectations of British effort, effectiveness, and loyalty, while being tolerably congruent with British national interests' (2008: 15). In material terms – notwithstanding for the moment normative commonalities – it has little choice but to do so; indeed, the point here is that the instrumental nature of the relationship may ultimately drive adherence to normative elements. The interplay between material and normative elements, however, suggests that Britain's national interests are not some fixed quantity standing in contradistinction to transnationalised defence policy, but are themselves malleable to the transnationalisation process.

Crucially for the purposes of this work, Edmunds' argument supports King's conclusion that it was not strategic logic that defined European and UK participation in Afghanistan, but the 'institutional dynamics' of interdependency and transnationalisation within NATO (King 2011: 25). Indeed, at the commencement of each major increase of ISAF and British activities in Afghanistan – namely, the expansion of ISAF's writ beyond Kabul in 2003 and the UK's deployment to Helmand in 2006 – one may find the dynamics of collective security obligation to be at least as important as theatre-specific strategic considerations (Bird and Marshall 2011: 154–5). Centrally, it is a contention of this work that it was not primarily the direct threat of terrorism in Afghanistan or weapons of mass destruction in Iraq that animated British military participation in those conflicts, but rather the direct imperative of satisfying American expectations and of contributing to collective security operations in general. Since the UK is dependent upon the US for the realisation of its own strategic objectives, it is natural that this would be the case, for without satisfying American expectations, very little is possible for the UK. Transnational pressures are observable at each point of change in Britain's Afghan strategy, from the original decision to 'stand shoulder to shoulder' with the United States in 2001, to the expansion of ISAF's role in Afghanistan in 2003 following the Bush Administration's invasion of Iraq, the deployment of British forces in Helmand in 2006 in response to Britain's foundering campaign in Basra province in Iraq, the expansion of British forces and change in operational approach following the Obama Administration's decision to employ 'population-centric counterinsurgency' operations in Afghanistan in 2009, and finally to the decision made by Prime Minister David Cameron in concert with Obama in 2011 to begin an accelerated drawdown of personnel in the country. Indeed, underscoring all of Britain's activity in Afghanistan was the 1999 Blair Doctrine, upon which Britain's involvement in the War on Terror was based. This doctrine represented a concise and unambiguous advocacy of transnational policy, linking as it did the assumptions of 'liberal peace theory' with a desire to equate Britain's national interest with those of the international community (Blair 1999, online). Taking this doctrine into account, it is plausible to conceive of Britain's participation in coalition stabilisation efforts as equivalent to the Blair Government putting its principles into practice.

When viewing British planning for Afghanistan along the lines of a core imperative of satisfying expectations, the transnationalisation of defence policy in these conflicts can be held up as a major factor leading to the overstretch of British military capabilities and as a primary cause of Britain's inability to redress this overstretch due to the constraining effect of transnationalisation on policy choices available to the British state (Edmunds 2010: 381–2). In making the case for the centrality of transnationalisation as both dictating and limiting UK options for policy and strategy in Afghanistan, it is perhaps useful to consider briefly a comparison between the UK's policy orientation and strategic options in the War in Afghanistan with those of the British Empire in the Second Afghan War of 1878–1881. The policy origins and strategic debates surrounding these conflicts have much in common, despite the obvious differences in technology and operational acumen that a time gap of over a century might produce. The Second Afghan War came about as a result of Imperial concerns over the stability of Afghanistan in the face of external (Russian) intrigue and the potential implications to the security of British India as a result of an unfriendly Afghan regime. The policy debate in Westminster focused on the nature of the threat posed by the loss of Afghanistan to the Russian sphere of influence in Central Asia and the most appropriate strategic posture to take in response. The ruling Conservative Government (under the Lord Beaconsfield, Benjamin Disraeli), advocated an aggressive 'forward' policy which involved setting up a British diplomatic mission in Kabul and the possibility of punitive measures against rebellious Afghans. The opposition Liberals, led by the Marquess of Hartington and the figurehead of William Gladstone, promoted instead a 'rear' policy – known at the time as 'masterly inactivity' – which centred on the use of subtle diplomatic bargaining guided by adherence to a normative code of non-interference in tribal affairs and the increased protection of the Durand Line separating Afghanistan from British India. Parallels between these positions and those of the present day are observable: as with the forward policy, the stabilisation approach was informed by a perceived need to implant international forces to bolster the position of an allied leader, whilst the counter-terrorism approach has argued for the avoidance of involvement in the political and social maelstrom of Afghanistan in favour of a more limited campaign of containment and suppression of insurgent activity within the borders of Afghanistan.

Again, it is worth restating that such parallels are naturally imprecise given the significant differences between the late Victorian period and today. Other variables are obviously important in distinguishing the two conflicts; indeed, much has been made of the difficulty of devising strategy and effective communication of strategy in an era of 'mediatisation' of conflict and a corollary multiplicity of political voices (Price 2009; Gilboa 2008; Gowing 2009; Brown 2003; DCDC 2012). Nik Gowing (2009: 11, 77) describes the information environment in which states jostle amongst non-state actors for communicative power as an 'almost infinite, digital landscape' where 'information doers' – meaning practically anyone with a camera phone – have contributed to the creation of 'a media spectrum and matrix that has rapidly become far wider, deeper, more multi-dimensional and all-pervasive' than anything previously known. States must be able to respond to crises as they happen, but the democratisation of the information environment also requires them to respond consistently and without contradiction between audiences. Messaging can be undermined by the reporting of countervailing realities, and trust between publics and institutions can be rapidly eroded. However, while the technological differences between the twenty-first and nineteenth centuries need no real elaboration, it is crucial to our understanding of what is most significant in the differences between these two campaigns to recognise that media and political issues presented considerable problems for the state in the nineteenth century as well. The domestic press was highly critical of the actions of the Conservative Government that led to the deployment of forces into Afghanistan throughout 1878 and even more so as the campaign drew to a close in 1879–1880. Specifically, it was only weeks prior to the announcement of the forward policy by Government (as a response to Russian advances in Afghanistan) that Disraeli had declared peace with the Russians at the Treaty of Berlin. News of Russian forays into Kabul therefore represented a political embarrassment for the Government and gave credence to the argument made by the liberal press that the forward policy was ill-thought out and ignorant of previous costly military encounters with the Afghans. Indeed, *The Times* – a stalwart supporter of the Government's policy – paraphrased the feeling of growing discontent amongst the fourth estate as follows:

> We have entered with more or less of disguise upon a policy of aggression, undefined in its aim, and certain to bring trouble with it whatever way it may end. We are now only at the commencement of the difficulties we have brought upon ourselves. (*The Times* 1878: 9)

During this period political debate in Westminster surrounding the efficacy of the forward policy was far more pronounced than any surrounding the War in Afghanistan, with clear splits between the main parties over the rightness of the mission. Liberal proponents of masterly inactivity warned against the potential for troop overstretch and strategic stalemate that would, in their minds, accompany a failure to stabilise Afghanistan through achieving a functional governance structure. In mid-December 1878, Earl Grey outlined this view in opining that,

> I do not mean to say that the Afghans will be able to drive you out of their country by force; no, but the troops you could maintain there would be so

> harassed and worn down by the continual hard work imposed upon them, the
> difficulties of governing the country would be so great, that in the end you would
> find it practically impossible to persevere in the attempt. (Hansard 1878)

By early 1880, a series of high-profile setbacks (such as the assassination of the British envoy, Sir Charles Cavagnari, the abdication of the pro-British Afghan Amir, Yakub Khan, and the recommencement of hostilities by the Afghans against British positions) led to a renewed round of criticism in the Commons. By this point, the debate had centred upon the overall aims and objectives of the war, with those opposed to the forward policy arguing that there was no certainty in who the British were fighting or why. In February 1880, Liberal Member William Harcourt issued a scathing riposte to the Conservative forward policy and its search for a 'scientific Frontier', stating:

> they wanted a barrier. A barrier against whom? Who was the unknown foe against
> whom we waged these mysterious wars, to baffle whom we attacked chieftains
> who were not our enemies, invaded countries with which we had no quarrel,
> incurred ruinous expenditure, experienced appalling disasters. (Hansard 1880)

By characterising the forward policy as foolish, naïve, disastrous and immoral, the Liberals in Westminster and the liberal press reaped political dividends. The issue of morality of the war was seized upon by them in the 1880 General Election, when former Prime Minister William Gladstone made criticism of Conservative foreign policy the centrepiece of his famous 'Midlothian Campaign', leading ultimately to the toppling of the Disraeli Ministry, a return to power for the Liberal Party and a reversion to masterly inactivity in Afghanistan.

Given that dissent against Government's Afghan policy was at least as intense in the nineteenth century as it has been in the twenty-first, and given that such dissent came from both political opposition and the media, it is difficult to maintain the position that the strategic issues faced by UK messaging from such quarters today is something wholly new or fundamental. On the contrary, the costs of intervention in this century can be seen in political terms as far less severe: unlike in the past, Governments have not been dispossessed of power because of their handling of the conflict. Indeed, the 2010 General Election's televised debates featured no real debate over Afghanistan, save for disagreements about operational requirements (BBC 2010: 5). In comparison with the tumult of the nineteenth century, the current political order in Britain has been remarkably unified. Rather, the most considerable difference between the Second Afghan War and the recent War in Afghanistan, as far as British strategy is concerned, is in the degree of autonomy the state has had in shaping its respective strategies. The Empire was for all intents a self-contained and self-reliant unit of power on the world stage; whilst there were undoubtedly machinations between it and other Great Powers, and within it, between London and its regional administrators, the constraints upon its strategic options were limited in comparison with today. On the one hand, autonomy of interests meant that policy and strategy could be meaningfully debated by domestic political parties. On the other, the state could operate with unity and with relative impunity in its military endeavours and, despite the deep moral misgivings many Liberals had with the forward policy, it was not limited by liberal norms of state behaviour as the contemporary state is. As such, where diplomacy and peace offerings failed, the Empire could (and frequently did) put down threats to its security through pursuit of punitive measures.

For example, Major General Frederick Roberts, the commander of British forces in Afghanistan in 1880, engaged in tactics of 'butcher and bolt' raids, property destruction and mass exiling of rebel forces to quell the insurgency, and exercised his authority to 'execute rebels who had defied the will of their Amir, broken diplomatic protocol and besmirched British prestige' (Johnson 2012: 1–2, 22). Such tactics are not entirely dissimilar to those proposed by contemporary advocates of a counter-terrorist stratagem in Afghanistan, but these are obviously largely untenable today because their use would breach the liberal normative basis upon which the collective security apparatus of NATO and ISAF is founded. Indeed, such an operational approach is more readily applicable to the US-dominated OEF's 'drone' policy. Therein lays another comparison: the United States has the resources and capabilities to undertake a counter-terrorist posture in Afghanistan because it can act independently of its alliance partners; it has relative autonomy in the same manner the Empire once did. The Empire's autonomy of action allowed for a relative independence of strategy and policy, and was therefore unaffected by the *quid pro quo* of alliance dynamics and transnationalised policy. As such, they were less prone to transnational dilemmas and a concordant perception of the need for strategic communication remedies. They did not think about how to communicate strategy; they simply 'did' strategy. The constraints of alliance dependency today, however, have produced transnationalised effects that have curtailed autonomy, divested authority, and contributed to strategic vacuums which have allowed for departmental agendas to manifest and conflict in a way that has severely affected British strategy in Afghanistan. In this way the communication dilemmas of the contemporary state are as much internally as externally located. Edmunds has outlined these tensions in a manner that can be subdivided into two themes: those concerned with the future role of the armed forces as a result of the drive toward transformation, and those between the various institutions and departments of the British state in response to the operational challenges of Afghanistan.

Intra-state Tensions

The first set of tensions has been described by Edmunds (2010) and Paul Cornish and Andrew Dorman (2009a) as reflecting a philosophical debate within the defence community and academia over the relative merits of two theses that take opposing views regarding the force projection consequences of military transformation. The first of these two perspectives, Rupert Smith's 'war amongst the people' thesis, posits a diminishing utility of conventional military force, the increased importance of counterinsurgency, and the centrality of 'influence' to military strategy and of 'hearts and minds' as the centre of gravity in the majority of contemporary 'low-intensity conflicts'. A real-world consequence of the Smithian view was the perceived necessity of prioritising land forces (essentially the Army and Royal Marines) over and above the Royal Navy and Royal Air Force (Smith 2006: 200). The second perspective, accorded in the main to Colin Gray, has been identified by Edmunds and Cornish and Dorman as a 'balanced' approach to transformation which, whilst conceding that there is currently a proclivity of 'small wars' and therefore the need to adjust operational capabilities to meet the demands such conflicts necessitate, warns against the creation of an imbalance between services to such an end, as well as the potential for 'presentism' (focusing on today's requirements at the expense of those of the unforeseen future) to skew Britain's medium term capabilities (Gray 2008: 17). Various scholars have since noted, by way of references to budgetary allocations (and budgetary cuts) and the direction of military doctrine, that the Smithian transformation agenda and its main proponents won the debate, and that the 'wars amongst the people' thesis – and its corresponding emphasis on 'effects-based' operations and population-centric methods of counterinsurgency – have been internalised by many within the Army as best representing its own future (Farrell and Gordon 2009: 679–80, Cornish and Dorman 2009a).

Along these lines, Cornish and Dorman argued in a series of articles for *International Affairs* between 2009 and 2012 that the British state's opting for the Smithian perspective on transformation resulted, in its implication that the material improvement of UK land warfare capabilities was central to British security interests, in severe inter-service rivalry due to the prioritisation of 'expeditionary land warfare' (and therefore the Army) above the Royal Navy and Royal Air Force (2009b: 738–9). For them, the reconfiguration of service budgets (latterly under conditions of austerity) and the apparent subordination of naval and air power to land power produced an environment of considerable enmity between the services, as each seeks to protect its own interests in what the authors have dubbed 'campaign tribalism' (Cornish and Dorman 2009a, 2009b, 2012: 215–17). They contend that this development stemmed from a long-standing decay of the relationship between policy and strategy emanating from a lack of civilian leadership on defence matters and a correlating deference to military leadership. The consequence of this, in their view, was that defence reviews were effectively 'defence-led' (2009a: 54, 2009b: 740–41). In addition to these institutional issues, the effects of austerity on such review processes have, for Cornish and Dorman, also exacerbated tribal behaviour between the services and a subsequent non-strategic or even 'anti-strategic' battle for pre-eminence (2009b: 739). Thus, by reference to the inter-service fallout that transformation has precipitated, we can begin to see how the obligation-cooperation cycle implicit in the transnationalisation thesis has produced severe implications for the cohesion of the British state.

The second set of intra-state tensions stems largely from the transformation agenda's role in influencing the internal divisions within the Armed Forces and the policy-strategy vacuum and budgetary issues resulting therefrom. Specifically, it relates to civil-military relations over resource provision and the balance of authority within the 'Comprehensive Approach' for Afghanistan between the Army and its civilian partners at DfID and the FCO. The often public debates around these issues not only exposed the fragile working relationship within Whitehall (and the challenges of fulfilling the baseline assumptions of the CA), they also laid bare how discord within the British state contributed to the operational and strategic difficulties faced by the British in Helmand. At the broadest level of analysis – looking at relations between Government and the Chiefs of Staff – an inference can be made that the decision to deploy to Helmand was the crucible upon which civil-military tensions were most severely tested. The motivations of both the civilian and military leadership in 2005–2006 were largely shaped by dynamics of transnationalisation: the Blair Government sought to maintain its support for the American Bush Administration by taking on a large share of the burden for ISAF's move south in Afghanistan, whilst some in the British military viewed Helmand as an 'opportunity for good soldiering' that would allow it to begin to extricate itself from its role in southern Iraq (Personal Interview 2013). In this fundamental way, both the civilian and military leadership were engaged in a kind of a-strategic appraisal of the significance of Helmand to their individual interests within the transnational milieu of British foreign and defence policy.

Assuming the plausibility of ideas of institutions as self-regarding, one can begin to make an assessment of how the overriding priority of living up to transnational obligations shaped what would prove to be an unrealistic plan for Helmand. As mentioned in the opening chapter, each element of the state involved in Afghanistan had their own preconceptions about what the ultimate purpose of the Helmand mission was and the correct ordering of methods in pursuing stabilisation. Along these lines, the planning process for Helmand was characterised by a former Government official as one of a coincidence of civilian and military interests, in which all parties agreed to the feasibility of the CA on the nominal basis that their own individual part to play would be given a place (Cavanagh 2012: 51). Indeed, in an extreme reading, one might conclude that the CA was 'comprehensive' for the institutional reason that it was the only way all viewpoints and interests could be accommodated simultaneously, and that for any approach to be viable it must incorporate the interests of all concerned parties. In this light, without making any sweeping indictment of its strategic practicability (for strategies that satisfy institutional interests do not necessarily always fail), the CA can be seen as a totem around which the divergent institutional interests of the British state could coalesce. Perhaps inevitably, then, when problems emerged almost immediately on the ground in Helmand that necessitated a review of the CA and the reordering of operational priorities, the recriminations between Government and Army that followed reverted to type of defending individual institutional interests at the expense of the state's overall reputation and, by proxy, public perception of the feasibility of the campaign.

The unravelling of the CA in Helmand in mid-2006 should therefore be seen as the catalyst for several years of institutional acrimony. A brief review of these tensions is instructive. The opening shots were provided in mid-2006 by the most high-profile actor in the entire affair, namely then-Chief of the General Staff Sir Richard Dannatt, who in October gave an interview with the *Daily Mail* with the intention of 'stand[ing] up for what is right for the Army' and wherein he criticised the Blair Government for its decision to force the Army to fight two wars simultaneously. He would later accuse Government in his autobiography of providing assurances to the military during the Helmand planning process that expansion in Afghanistan would correspond with a reduction of commitments in Iraq and a correction to what he saw as the general underfunding of the Armed Forces (Sands 2006, online; Hollander 2010, online; Dannatt 2011: 406). Indeed, some evidence supports his claim, including former Defence Secretary Geoff Hoon's view that simultaneous missions would be untenable and that reducing forces in Iraq would be necessary for expansion in Helmand (Iraq Inquiry 2010a, online). However, John Reid, the Defence Secretary during the Helmand planning process and the first months of deployment, refuted Dannatt's version of events at the Iraq Inquiry in February 2010 by providing documentary evidence showing that the then-Chief of the Defence Staff, General Michael Walker, had informed him in October 2005 that, in his professional view, the Armed Forces were indeed capable of performing concurrently in Iraq and Afghanistan (Iraq Inquiry 2010a, online). Along such lines, the evidence presented at the Inquiry appears to confirm the view that both sides were content with the leveraging of the Armed Forces to fulfil Britain's collective security commitments. Again, this is because doing so apparently fulfilled the interests of both senior politicians and military officials. Both Gordon Brown's (in 2005 Chancellor of the Exchequer) and Dannatt's testimony appeared – perhaps surprisingly given their apparent mutual animosity – somewhat congruent in their recollection that they and others involved in planning the Helmand deployment were aware that maintaining Britain's commitments to Iraq would mean that forces would be 'stretched but not overstretched' and 'running hot' (Iraq Inquiry 2010b, 2010c, online).

Despite the hostility of Dannatt and elements of the conservative press to Blair's war in Iraq, the public breakdown of civil-military relations plumbed new depths under Gordon Brown's premiership. By most accounts, Brown was never held in particularly high regard by the military, who saw him as unsympathetic to the demands of the Armed Forces and lacking knowledge in defence issues, and his Government in general as more concerned with (and more fiscally generous toward) domestic issues such as the National Health Service, education and social welfare (Bower 2007: 388; Chin 2009: 134–7). An example of this apparent lack of respect can be found in the reaction of some in the centre-right press and the Shadow Defence Minister, Liam Fox, to the appointment of Des Browne to Secretary of Defence, who not only was responsible for the financing estimates for deployment to Helmand as Chief Secretary to the Treasury under Brown, but also took on his role at Defence whilst simultaneously carrying out responsibilities as Minister for

Scotland (Coughlin 2008b: 25; Walker 2007, online). It was in this environment, where the Armed Forces perceived a lack of interest or a failure of leadership in issues of defence from the Brown Government that, according to Betz and Cormack, the military and MOD took the lead on Afghan strategy (2009: 327, 334).

Michael J. Williams (2011: 67) has argued that this state of affairs produced a view within the FCO and DfID that the MOD was dominating the CA and was more widely leading defence policy in the place of Downing Street. It has subsequently been argued that, as a result of both unforeseen challenges on the ground in Helmand and the lack of an ostensibly neutral leading role at the prime ministerial level to arbitrate between the institutional interests of these departments, institutional 'turf wars' developed over jurisdiction and control of Afghan policy and resources (Baumann 2009: 13). Although encompassing conflicts of interest across the spectrum of diplomatic, military and development agencies involved in Helmand, perhaps the best documented and most significant of these turf wars was between the Army and MOD, on the one side, and DfID on the other. The conflict between them appears to be long-standing (practically from the outset of British involvement in Afghanistan) and deep-rooted. To illustrate, in the first year of the Afghan intervention then-DfID Secretary Clare Short refused 'to allow DfID to cooperate with the MOD and FCO in developing Phase IV' stabilisation operations on the basis that, in her own words, the advice of DfID on substantive areas of development policy were 'ignored by government, who favoured supporting US counter-terrorism policy' (Chin 2009: 132; Short in Sedra 2010: 13).

This initial friction likely set the trend for a paucity of inter-departmental cooperation and the inadequate development of the Helmand plan (Chin 2009: 132). DfID and the FCO's approach to stabilisation, centred on soft power and development, did not correspond with the highly kinetic reality of Helmand in mid-2006, just as 16 Air Assault Brigade and its successors' approach to stabilisation jarred with the CA plan set out by Government, insofar as its command opted for dispersal and offensive operations over concentration and reconstruction. Faced with a highly kinetic combat environment where all development work would be highly dangerous and potentially deleteriously linked to military activity, the civilian departments were reticent to work with the military and MOD. In DfID's view in particular, reconstruction and development work was seen as untenable chiefly because their remit was 'post-conflict' stabilisation (epitomised by the moniker of the inter-departmental (FCO, DfID and MOD) team established for Helmand, the *Post-Conflict* Reconstruction Unit (PCRU)), whereas the situation in Helmand was very much in the 'conflict' stage of events (Betz and Cormack 2009: 326; Jackson 2010: 119; Lamb 2006: 22). According to an article in *The Times* in mid-2006, DfID's reluctance to work alongside the military in such an environment was given short shrift by the MOD, whose source argued to the press that if DfID were unwilling to work under the conditions as they were, their funding and resources should be reallocated directly to the MOD (Lamb 2006: 22).

In light of these tensions, one may see the case of the PCRU as a microcosm of the departmental turf wars at play in Helmand and as a case study of the gulf between the optimistic thinking of the planning process and the harsh realities of implementing the CA, both in terms of how it played out on the ground and in terms of how its delivery was pursued by self-interested institutions. According to Williams, the advent of the PCRU in Helmand was met with hostility and 'obstructionist' behaviour from all three departments (as well as the British Embassy in Kabul) as representatives from each were 'worried that the PCRU was forming new relationships and would upset their own existing relationships' in Helmand (2011: 72). In other words, while the PCRU was the elected means of Government and its constituent parts for delivering the CA in Helmand, the inter-departmental turf wars brought to the surface by its creation and implementation effectively impeded its ability to function properly, suggesting that the CA itself was either unrealistic in formulation or that it failed to adequately account for the extent of inter-departmental differences. Another possibility, to synthesise these arguments, is that it was unrealistic in part *because* it functioned to fulfil all these interests simultaneously. Whatever the case, from a military perspective it was held that this 'disconnect between the various arms of government has undermined faith in the Comprehensive Approach and whether it can be delivered', leading one senior British military commander, then-Brigadier Andrew Mackay, to conclude that the CA was 'seen [by the military] as a Whitehall concept which had no actual impact on the ground in Helmand' (Farrell 2008: 795). The problem the CA raised was that it sought by its very definition to subordinate the military role under a wider set of non-military goals (Edmunds 2010: 390), but in doing so left the military in the position of carrying the greatest burden whilst remaining beholden to the demands of civilian agencies that were unwilling or unable to carry out their own responsibilities.

As referenced earlier, all of these issues stemmed from a divergence of opinion between the civilian agencies and the military of the appropriateness of the CA's core assumption of 'securitising' development (Duffield 2001). Whereas the military and the MOD (the authors of the CA) could see this approach to development as logical and necessary because it reflected their central institutional interest of pursuing national security interests, for DfID and the FCO their core objective of reducing poverty was relegated to a means to the end of security (Farrell and Gordon 2009: 680; Howell 2010, online). Transnationalisation is relevant here as one may make the inference, as Clare Short did, that the securitisation of development in Afghanistan was a consequence of the UK's interest in following an American policy in Afghanistan that has been dominated by national security concerns over and above developmental ends. If one accepts the provenance of Short's point of view and the core arguments of King (that multinationalisation has led to transnationalisation) and Edmunds (that transnationalisation creates or heightens intra-state tensions), it is reasonable to conclude that the conflicts of interest hitherto outlined are the by-product of Britain's participation in collective security frameworks in Afghanistan. From this point of departure, one is in a position to claim that much of the operational and strategic challenges of the UK regarding Afghanistan were internally located, albeit (crucially) externally precipitated. Moreover, the lack of a unified political objective within the British state allows us to account for some of the strategic incoherence in Helmand as well as the incompatibility of the institution-specific policies of stabilisation, counter-narcotics and counter-terrorism.

The UK's 'Transnational Dilemma' for Afghanistan

The existence of multiple incompatible policies for Afghanistan is, therefore, established as the result of the intra-state tensions produced from the UK's transnational policy posture and a concomitant lack of strategic direction and autonomy at the national level. How these policies unfolded over time (as 'policy narratives') can be explained in the context of such tensions, with ultimate reference back to Edmunds' transnationalisation argument. The story expounded in the empirical chapters of this work explains how the three policy narratives employed by the British state regarding Afghanistan originated from a milieu of institutional interests interacting under the constraints of a transnationalised policy environment which, because they reflected those interests and not the strategic requirements of the Afghan mission, failed to coalesce into a workable whole. As Britain's military (and therefore political) commitments to Afghanistan increased in 2006 with deployment to Helmand, the shortcomings of this arrangement – epitomised by the early failure of the CA – became obvious. Britain was expending substantial human and material resources in an undertaking that did not appear to have any unified purpose, primarily because it had multiple purposes depending on whom one asked. According to the transnationalisation thesis, if there was an ultimate purpose for the British state, it would be that of supporting American defence policy as a dependent mid-level power, and this imperative would therefore have taken strategic precedence over the requirements of stabilising Afghanistan, which would amount to a means to an end. The notion that Britain's decision to stand 'shoulder to shoulder' with the United States has defined its participation in the War on Terror, and the fact that the last two National Security Strategies and the 2010 SDSR (amongst other documents) have explicitly stated the centrality of the special relationship to Britain's national interests (security-related and otherwise) should make this evident enough.

Balancing national and transnational interests was a difficult task for the UK in Afghanistan because British 'national interests' are uncertain, due to the inter-related facts that they are (a) largely shaped by and contingent upon US interests, and (b) a composite of the various institutional interests (themselves shaped by the demands of transnationalism) within the state itself. Since Britain's national interest of territorial security is, according to its own strategic documents, practically impossible to secure on its own in the manner it deems necessary, it is difficult to conceive of defining the national interest in isolation of the interest of its security guarantor. Similarly, the notion of a national interest (or perhaps more accurately a state interest (*raison d'état*)) can only be understood in practice as the lowest common denominator of the interests of its constituent parts, and beyond that there are issues of interpretation to contend with regarding how interests are best secured. As the previous discussion has shown, British state institutions in Afghanistan have often interpreted the national interest in conflicting ways.

These issues contribute to a transnational dilemma between domestic politics and international diplomacy, where interventions are indirectly linked to the interests of the UK polity but from which the British state 'cannot exempt [itself] for fear of being excluded from access to critical shared security goods' (King 2010: 388). What this implies, first and foremost, is that transnational *realpolitik* is at least as much a motivation for the UK's involvement in Afghanistan as concerns about terrorist plots against its territory or overseas interests, but also that interests must be made 'saleable' to the general public in terms that generate popular support. This is, of course, an unstatable reality for any self-regarding politician in any liberal democracy (since the implication would be that the lives of British soldiers were sacrificed in the pursuit of maintaining American favour rather than for vital and direct security reasons), and therefore encapsulates the communication element of the transnational dilemma, where public acknowledgement of the dilemma must be avoided at all cost lest it unravel the fabric of Britain's transnationalised interests and policies. Thus King is undoubtedly correct in his contention that states 'have found it very difficult to conceptualize this transnational dilemma or explain it to their publics' (2010b: 389), not only because it would be entirely self-defeating (in terms of a national interest contingent upon the provision of collective security goods) to do so, but also because it would fail to accord with the sensibilities of liberal democratic politics.

Instead, the solution for Afghanistan was to simply ignore the transnational dilemma. British politicians tended to emphasise one of two positions on Afghanistan: the first, running from 2001 until 2007–2008, related to the security of the international community at large and the well-being of Afghans and the democratisation of their society. The second, from 2008 to the present day, has emphasised the mission in Afghanistan in terms of its relationship to British security interests, increasingly at the expense of any humanitarian or international motivations. The fact that 'national' security reasons were not frequently referenced until this point suggests that they were not of paramount importance in Government's decision to take part in the war in Afghanistan, that 'nationally' oriented foreign and defence policy was not the lens through which security issues were conceptualised by Government or, at the very least, that Government felt such concerns were not necessary to justify British participation. The use of such arguments only became reasonably regular in occurrence from late 2008, presumably as a reaction by ministers to the increasing numbers of casualties and a corollary increase in media attention on the Helmand campaign.

Rather, prior to October 2008 (and in a few isolated cases after that date), communications by ministers and state officials regarding Afghanistan tended to follow a pattern of deploying rationales for the intervention that reflected their own departmental briefs in conjunction with the emphasis on democratisation coming down from the prime ministerial level. Indeed, according to Dannatt, the Labour Government had placed an embargo on military commanders issuing their own statements (which invariably would have reflected the kinetic aspects of the campaign) and instead enforced a 'pre-ordained narrative' of reconstruction and development (Dannatt 2011: 323, 354). Dannatt's assertion of narrative scripts can be easily validated: DfID Secretaries such as Hilary Benn and Douglas Alexander, for example, cited improving the lives of Afghan women or the education of young girls as a reason for intervening in Afghanistan in a way that placed such efforts on a par with security concerns (Benn 2006; Alexander 2008). Defence Secretaries such as John Reid and Des Browne would emphasise reconstruction and counterinsurgency (with Reid in particular fixated on the role of counter-narcotics as compatible with the latter) even when those elements of the mission were practically non-existent in comparison with kinetic activity. Perhaps most remarkably, Home Office minister Phil Woollas, whose brief related to border control, argued in 2009 that Britain's presence in Afghanistan helped prevent mass immigration from that country to the UK (Slack 2009, online). In these cases, the issue was not one of the veracity of each claim about the significance of Afghanistan – each issue was of importance to some collection of ministers or civil servants and their respective role in Government – but rather one of consistency of message, and who controlled the message.

Who controls the message, and indeed whether messages are controlled enough to constitute 'the message' (as opposed to multiple messages) has real consequences for the viability and popular support of a policy and strategy. Along such lines, Betz and Cormack (2009: 329) offered an explanation that accounts for many of the consequences of institutional turf wars and departmental fragmentation of messaging, pointing to the fallout of public spats over equipment shortages, the unpopularity of counter-narcotics efforts, increasing troop casualties, the negative association of Afghanistan with Iraq, and a general feeling of strategic drift as contributing factors. Indeed, the first two and the last of these variables can be related directly back to turf wars stemming from institutional tensions associated with the strategic void concomitant with transnationalised policy, while the third and fourth factors (the so-called 'poodleism' effect) can be indirectly associated with Britain's transnational positioning of foreign and defence policy. The notion that Britain was carrying a disproportionate burden in collective security operations characterised by European 'free riders' – and by extrapolation that the obligation implicit in Edmunds' transnationalisation thesis was inducing negative consequences for the British state – is supported by troop casualty statistics: by 2007, Britain's military deaths accounted for 40 per cent of the total for all European contributing nations, and by the end of 2011 was greater than those of all European nations put together (iCasualties, online).

The combination of these factors over the first 18 months of the Helmand campaign may be seen as precipitating factors in causing the Brown Government to attempt to address its shortcomings in developing a coherent and consistent set of political messages for the UK public on Afghanistan. The prevalence of political discussion on the inconsistency of messaging on the conflict by late 2007 suggests that the dilemmas that Britain's Afghan policy (and its transnational orientation more broadly) had produced could no longer be ignored. Pre-2006, the costs of collective security obligations in Afghanistan were minimal and largely overshadowed by events in Iraq. Following the deployment to Helmand, however, the dilemma of balancing the costs of participation against the benefits became one that was no longer politically tenable or explainable by reference to democratisation and counter-narcotics aims. Because mixed messaging was the result of 'information fratricide' between competing Government departments, streamlining messaging to avoid public confusion about the purpose of Britain's presence in Afghanistan meant getting to grips with the turf wars between departments of state. It was in this context that Government began to adopt a more self-aware and reflexive approach to public relations on Afghanistan which would later be referred to as 'strategic communication', developing doctrine and institutional processes with the aim of prioritising a key message that could unify all efforts in the CA and all elements of state.

Strategic Communication: A Solution to or Continuation of the Transnational Dilemma?

The origins of SC as an institutionalised process and appendage of MOD bureaucracy can be traced back to early 2009 (see Kirkup 2009, online). This is chronologically significant because it coincided with the beginning of the reconfiguration of the British narrative for Afghanistan by Labour Defence Secretary John Hutton in October 2008 along lines that would mirror the MOD's definition for SC. The contention here is that Hutton's reconfiguration of narrative and the development of SC are institutionally, chronologically and thematically linked, suggesting that they were, in fact, two parts of a larger development within defence aimed at reconfiguring the Afghan narrative. In his maiden speech as Defence Secretary, given to the International Institute for Strategic Studies in November 2008, Hutton argued for the significance of Afghanistan in predominantly military terms, stating that the successful prosecution of the campaign was a 'vital national interest' and, in the process, coined a new term – 'national security interest' – to describe the significance of Afghanistan. Interestingly, this term has a way of addressing the transnational dilemma without committing too much rhetorical weight to either of its constituent elements; its strength was in its subtlety. It stated outright neither the necessity of meeting collective security obligations nor the directness of the threat posed by Afghan terrorism to the UK mainland; rather, it occupied a medium space that hinted at both and, in doing so, made it possible to bridge the national with the transnational without ever referencing the substantive meaning of their relationship. It positioned UK policy on Afghanistan in a space where it could posit the necessity of collective security operations there but without exposing it to the deficiencies associated with collective security, of which there were many. Indeed, Hutton's speeches contained several references to UK displeasure at free riding problems and the damage it was doing to NATO's credibility (Hutton 2008b). More broadly, Hutton's speech ushered in a post-Blairite framework for defence policy rhetoric, placing ideas of 'the national interest' and the primacy of 'national security' over liberal interventionist norms of democratisation and humanitarianism. From November 2008 onwards, a consensus amongst state officials – Government and Opposition, ministers and civil servants alike – would develop (evidenced by their public utterings) that would cement Hutton's reconfiguration.

It was in this context that MOD SC practices and processes developed. Unlike his predecessor Des Browne, Hutton seemed to have had the confidence of the Armed Forces, to the extent that, according to Seldon and Lodge (2011: 207), he was seen by some in Cabinet as the 'military's voice' in Government. This was reflected in a series of interviews and public statements in late 2008 and early 2009 in which Hutton sought both to reframe the Helmand campaign as one that was centrally a military one, and to downplay the developmental and diplomatic aspects of the CA pushed by DfID and the FCO. Specifically, Hutton's emphasis was on the centrality of military force in order to have any prospect of securing political settlements or stability for development projects. He repeatedly emphasised force where his predecessors were reticent to do so. In this way, Hutton's publicly stated view that military means could be effective in providing the space for a resolution to the conflict jarred slightly with those of David Miliband and others at the FCO, who spoke frequently of the need for a 'civilian surge' and the impossibility of a 'military solution' to the conflict (Miliband 2008a, 2008b, 2008c). Of course, an institutional perspective suggests it is perhaps natural that Hutton would emphasise the military dimension of the Afghan conflict given his departmental remit as head of the MOD (not to mention the operational requirements of the day of succeeding in 'clear, hold, build'). However, it is noteworthy that focusing on Afghanistan as a *national security* issue and fundamentally a *counter-terrorist* mission would have been difficult to substantiate without an emphasis on military prosecution on the ground. Indeed, anything less than a 'national security'-informed approach would have suggested the continued prioritisation of developmental efforts. Likewise, the statements of Hutton and other fellow travellers of a national security approach, such as Liam Fox, indicate that the view of some in Government and Opposition at this time was that an increased role for military force would be a far more difficult sell to the British public if the overall mission remained framed as a democratisation mission or concerned primarily with counter-narcotics work. In short, Hutton's reframing of the mission's purpose fit naturally with an emboldening of military rhetoric.

Press reports in early 2009 confirm that these developments coincided with a ratcheting up of institutional effort in SC, with articles noting the Cabinet Office's opening up of positions for strategic communication officers to direct Government messaging efforts (Walker 2009: 8). The actual theoretical and practical content of SC efforts seems to have originated mainly from the MOD, however. In February 2009, James Kirkup of *The Telegraph* appeared to confirm that Hutton's shift in narrative reflected fears within the MOD about the public's growing perception of transnationalisation of defence policy as animating Britain's motives in Afghanistan; he claimed that 'the MoD is haunted by the echoes of Iraq, and the belief among some voters – and MPs and soldiers, come to that – that Tony Blair ultimately sent UK forces into combat to preserve British relations with the US … [t]his strategy dictates that any decision to send more British troops must be explained in terms of the immediate UK national interest, not esoteric diplomatic considerations: we're not doing this to please the guy in the White House, we're doing this to keep you safe' (Kirkup 2009, online). The transition of the Afghan 'narrative' from an internationalist, humanitarian and democratisation mission to one concerned almost solely with British national security can be seen in this light as a direct response to concerns that the transnational dilemma had emerged into public discourse. Indeed, Kirkup's article represented exactly that; according to him, the worst fear of the MOD was the public uncovering of transnationalised defence policy and, as a consequence, the popularisation of a conceptualisation of the transnational dilemma in public discourse. Awareness of the shaping effect of transnational issues on UK policy and strategy would potentially open up state discourses to questions of inter-state bargaining and the inherent indirectness of the threat from Afghanistan. Needless to say, this is a highly nuanced account to give and is therefore vulnerable to misinterpretation or distortion. In any event, the official SC response was to avoid this account altogether, with the MOD opting instead to frame British efforts in Afghanistan as directly related to UK security without reference to any transnational features. The UK appears to have acted alone in this respect: according to an anonymous NATO official, few of the UK's European allies shared Hutton's conviction that Afghanistan was vital to their own security, contending instead that 'some of them believe the more fighting there is in Afghanistan, the less secure they are at home here in Europe' (*The Guardian* 2009).

All of these developments appear to have been motivated both directly and indirectly by the transnationalisation of defence policy, as an attempt to both pre-empt and remain relevant to American strategy for Afghanistan, and to maintain a justification that could sustain British public support for transnationalised policy in the face of an upsurge in operations and casualties. The shift to the national security 'narrative' took place against the backdrop of a renewed interest in Afghanistan on the part of the new US President, Barack Obama, whose inauguration occurred within days of the first press reports on SC and MOD narrative reconfiguration. Indeed, amongst Obama's first actions as President was a call for America's European allies to commit more resources to Afghanistan, signalling a likely increase in the size and intensity of military efforts there. In 2009 the Brown Government would oblige the US by increasing British troop levels to their apex of 9,500; UK forces would suffer more than twice as many casualties (108) in 2009 than in 2008 (51). This sharp rise in blood price was met with a more concerted SC effort to frame Afghanistan as a national security mission throughout the summer of 2009, with ministers deploying new phrases to describe the ostensibly existential significance of the terrorist threat there (labelling Afghanistan as the 'crucible', 'axis', 'epicentre' and 'incubator' of terror) and an associated rise in occurrence of powerful, hypothetical assertions about what would happen if Britain and its partners withdrew prematurely from the conflict. At the same time, mentions of DfID and FCO counter-narcotics and development work all but disappeared from the narrative. Perhaps reflecting British awareness of Obama's lack of interest in the democratisation agenda of his predecessor, the reconfigured Afghan narrative explicitly separated security aims from non-security aims such as the empowerment of women and the education of girls. This was so to the extent that, by July 2009, the champion of soft power in Government, David Miliband, spoke in terms that downplayed development by stating that the UK was 'not in Afghanistan militarily because girls were not allowed to go to school' (Miliband 2009c, online).

Thus, the emergence of SC practices on the one hand and political discourses of national security interests on the other should be seen as two inter-connected elements comprising the British state's response to the transnational defence dilemma on Afghanistan. This historical development has at least two identifiable implications that shape the analyses of policy narratives in this work. The first relates specifically to the transnational dilemma, and is that one can see SC as largely fulfilling its immediate purpose for Afghanistan, in that it did manage to produce a consistent message based on the centrality of national security out of an institutionally delineated morass of contradicting aims. This was so even though, as Betz and Cormack (2009: 328) have claimed, the existential content of the message was not commensurate with the 'inadequate' commitment of resources by Government to the war effort. SC practices effectively reframed the narrative during a difficult period to defend the Afghan mission by appealing to *raison d'état*, thereby providing the state with breathing space to fulfil its collective security obligations and maintain its transnational policy posture whilst promoting a more culturally acceptable and nationally located rationale for British sacrifices that would, incidentally, not be compromised by the free riding of its ISAF allies. In doing so, SC practices effectively circumvented the transnational dilemma by couching UK collective security obligations in primarily national terms. SC also provided options to the Government by allowing it to focus on strategic and operational aspects of the campaign that supported (and were supported by) the national security narrative. The less-than-successful development and governance work could be downplayed as non-essential to British national security, whilst the more practical work of training the ANSF and providing a security umbrella for 'Afghanisation' could be pointed to as both relatively fruitful and as essential for stabilisation and, by proxy, British security needs. As such, the claim to be working towards the 'national interest' can be seen as having a constitutive force on the ground in Afghanistan, insofar as it provided a discursive exit strategy from all elements of the CA that were superfluous to the baseline of national security.

The second implication for the analysis of UK policy narratives on Afghanistan relates to the issue of institutional tensions within the British state. SC came at the price of marginalising the agendas of DfID and the FCO in favour of the national interest and security approach of the MOD. Curiously, unlike SC practices in the United States which have been instituted as a cross-departmental process, British SC doctrine has been formulated within the MOD's think tank, the Developments, Concepts and Doctrine Centre (DCDC), culminating in the publication of Joint Doctrine Notes 1/11 and 1/12 in 2011 and 2012. This led some commentators to conclude that British SC appeared to having something of a 'tail wagging the dog' quality to it, that is, that principles of state-wide SC were defined by the defence establishment rather than by Downing Street, Cabinet, or the recently formed National Security Council (Cornish, Lindley-French and Yorke 2011: 5). Given the rather drastic reorientation of the Afghan narrative from one that attempted (perhaps unsuccessfully) to accommodate all institutional interests in the CA under an equally comprehensive (if somewhat confusing) über-narrative for Afghanistan, it is not unreasonable to posit that the location of SC within the MOD allowed defence matters to set the agenda for British efforts in Afghanistan from 2009 to 2014.

Along such lines, it is surely debatable whether SC practice has truly overcome the institutional turf wars that benighted British strategic practice in Helmand. If anything, SC appears to fit into the trend of institutional competition documented in the preceding pages of this work, insofar as it effectively won the day for the defence establishment in relation to the other, civilian departments of state. The "poverty reduction" agenda of the FCO and DfID was demoted – rhetorically at least – from being of equal importance to (or essential to) security aims to being an unnecessary and impractical diversion from national security objectives. Meanwhile, the ascendancy of 'national interest' and 'national security' in defence and foreign policy discourse is such that it has now become a staple phrase in practically every political communique, from National Security Strategies and Strategic Defence Reviews down to the day-to-day parliamentary statement covering most domestic and foreign policy issues. The influence of these terms – and therefore the shaping power of the defence community on the discursive content of Britain's transnationalised policy outlook – seems greater than at any point since the War on Terror began. SC can also be interpreted as an extension of institutional turf wars within the defence establishment insofar as it has promoted expeditionary campaigns such as that of Afghanistan from the level of 'discretionary conflicts' to that of near-existential struggles for national security. In doing so, it promoted the Army *vis-à-vis* other services by framing such conflicts as 'vital national security interests', thereby consolidating the 'wars amongst the people' thesis and the prioritisation of land forces over the more conventionally oriented Royal Navy and Royal Air Force.

Conclusion

This chapter has argued for viewing Britain's operational and strategic challenges in Afghanistan as largely informed by the fragmenting effects of transnationalisation on the British state. Transnational considerations are at the heart of British foreign and defence policy because the UK depends on mechanisms of collective security for the furtherance of its national interests. Because the national interest is only realisable through such mechanisms, British policymakers have quite naturally opted for policies and strategies that strengthen the UK's position in the transatlantic alliance system. Taking such positions confers significant political and diplomatic benefits to the UK, not least of which being its ability to continue 'punching above its weight' in world affairs. Costs are also evident, however, and the severe toll taken by Britain's Armed Forces in Afghanistan is a case in point. The UK's role in Helmand is indicative of the iterative effect of political obligation and military cooperation, whereby its pursuit of its national interest through compliance in collective security missions deepens its transnational outlook. This trend has had the dual effect of locking Britain into the collective security policy for Afghanistan whilst simultaneously limiting its ability to explain the purpose of Afghanistan in national terms.

This 'transnational dilemma' has been compounded by the destabilising effects of transnationalisation's twin pillars of transformation and obligation on the constituent parts of the British state. Under the weight of increased pressure to adapt, cooperate and perform in an environment of immense strategic and operational complexity, state institutions have often instead competed with one another, struggling inside a vacuum of externally imposed policy and strategy to assert their own vision of the purpose of intervention in Afghanistan. This fragmentation resulted in the production of multiple and often incompatible 'policy narratives' for the Afghan campaign – those of counter-terrorism, stabilisation, and counter-narcotics, respectively – which tended to reflect institutional interests and, in doing so, threatened to expose the calculus of *realpolitik* underpinning the transnational dilemma. From 2009 onward, Government and Defence attempted to partially rectify the disjointedness of official discourse on Afghanistan by imposing SC practices, centred on a core ideal of the 'national security interest', in order to provide consistency of messaging within the state and, I have argued, to attempt to circumvent the unstatable logic of the transnational dilemma in a manner that adequately accounts for and balances Britain's transnational and national obligations.

The limit of SC is that it does not resolve the unanswered strategic questions that transnationalisation imposes: as a communicative tool, it cannot provide insight on the location of interests or the proper relationship between policy and strategy, nor does it overcome substantively the task of explaining the necessity of collective security and the significance of indirect security threats to the UK. In fact, in practice it has made a point – by focusing on the direct threat of terrorism to UK national security – to simply ignore these aspects altogether. This indicates another problem, namely that the centrality of transnationalisation processes to UK strategic practice suggests that it is doubtful whether ideas of 'national interest' can be a truly useful means of measuring risk in collective security operations. UK interests are operationally, politically and doctrinally dependent upon the maintenance of its alliances. Its interests are as much located in the maintenance of structures that further or allow its interests to be pursued than in any objectively-defined security issue. As such, any hope of producing a clear framework for measuring the validity of a strategy in terms of its relationship to policy is mitigated by the abiding reality that much of British policy is, in fact, derivative of and therefore contingent upon the maintenance and protection of collective (NATO and American) policy. Thus, any analysis of the 'national' aspect of interest immediately comes up against the problem of defining where 'transnational' aspects end and national ones begin. Yet, because the abiding feature of transnationalisation is its 'unstatability', there cannot be an open discussion about the relationship between collective and national interests. This is unhelpful for British strategic thinking; indeed, it is potentially crippling so long as notions of British 'grand strategy' persist. As an institutional response to this dilemma, SC has simultaneously contributed to setting a new foundational unit of analysis (national interest) for defence whilst also putting up discursive barriers that are designed to insulate and obscure the transnational reality and the institutional and normative parameters therein that both inform and conceptually confound ideas of the 'national'. In this reading, SC is a largely superficial plaster covering a gaping wound, and one that may ultimately cause more harm than good.

Thus, what follows in Chapters 3–6 seeks to posit an understanding of SC as a response to a failure of strategic thinking at the national and transnational level. This contrasts with most academic and military doctrinal contributions to the subject, where SC and the institutional issues affecting the development of coherent stabilisation strategy have often been treated as separate phenomena. I propose that analysis of the relationship between policy, strategy and narrative as it unfolded in Afghanistan allows us to link the development of SC to issues of institutional discordance arising from transnationalised policy. This provides us with a new way of looking at SC and the relationship between unstated transnational policy and stated national policy. The empirical chapters of this work explain in greater detail how SC arose as a response to strategic challenges in Afghanistan, but frames it as a consequence and continuation of the transnational dilemma rather than as a solution to that dilemma and corresponding quandaries of national strategy.

Chapter 3
The Rise of the Stabilisation Narrative

It is necessary to preface the empirical chapters of this work with a brief discussion regarding the relationship between 'policy' and 'narrative'. In the following chapters these terms are used interchangeably or simultaneously, but it is important to recall at the beginning that they mean different things. Simply put, policy is the chosen path of a government, an initial statement of intent and a framework for the ways, means and ends of strategy, and should act as a guiding light for operational activity. Narrative, on the other hand, should be understood as how policy plays itself out over time in the form of a story, and contains all the elements of a 'storyboard' or 'script', including rationales, justifications, counter-arguments, rhetorical devices, and cultural and personal idiosyncrasies of a multiplicity of narrators. Policies inform narratives insofar as they are the starting point from which a narrative takes shape and evolves, but because narratives are the unfolding of policies over time, they also inevitably impinge upon the static nature of policy by breathing life into it and serving as the means by which policies respond to unfolding events that affect its coherence and validity. In order words, while policies determine narratives in their point of origin, they do not wholly constitute them. Thus, it is possible to speak of 'policy narratives' as all that is said about and in defence of a policy following the point at which the policy is first put into practice. It is also important to distinguish 'policy narratives' from 'strategic narratives' in the sense that narratives are not always 'strategic', a point the following chapters will repeatedly make clear. As the purpose of the following chapters is to chart and analyse the evolution of policy narratives as they unfold against the exigencies of the transnational dilemma, I have consciously taken the step of treating all articulations of policy as contributing to the overall narrative of that policy: a 'policy narrative'.

In each of the following four chapters, I argue that the policies that informed subsequent policy narratives became increasingly untenable over time as a result of the constraining features of the collective security framework (and the liberal norms underpinning it) from which they were originally conceived. The abiding realities of the transnational dilemma that the British state compels itself to negotiate has meant that its ability to direct strategy in Afghanistan has been distinctly limited and superficial, subordinated as it was to the direction of the United States as senior partner in the NATO alliance. In all three policies, this lack of strategic independence resulted in a loss of strategic clarity over time as to the purpose, appropriateness, and mutual compatibility of policy goals. This is turn came to be reflected in evermore disjointed policy narratives aimed at justifying increasingly incompatible policy objectives, and ultimately led to a recognition by the British state in late 2008 of the need for a narrative 'reset'. As I have already argued, this reset took the form of a self-conscious or reflexive attempt to reframe the 'genre' of the policy narratives to one of 'national security interests' in order to recapture dwindling public support for the Afghan mission, by placing the mission in traditional realist terms that could be apprehended by the public (and state officials) more easily than the imprecise and complicated language of collective security obligations. Central to this effort was the role of SC as a doctrinal and institutional implementation of codes of best practice in order to convince audiences of the 'national' necessity of continuing military activity in Afghanistan.

In this sense, the story of each of the three policy narratives analysed in the following chapters is one of the British state struggling to assert its control, not over the material aspects of strategy – which were largely beyond its control – but instead over the way in which these policies (and the strategies that were implemented to the end of those policies) were presented to the British public. This, I argue, was the case for the same reason policymakers now speak of narratives as 'strategic': because, to put it plainly, the nominal control British policymakers had over the way in which strategies were articulated was one of the few aspects of state power still largely in the hands of the United Kingdom. These empirical chapters demonstrate how the evolution of Britain's Afghan policies indicate the constraints of the transnational dilemma on the original narratives of counter-terrorism, stabilisation and counter-narcotics, and how this informed both the introduction of SC practices and the conceptualisation of narratives as 'strategic'. Transnational demands and correlating institutional tensions explain how the stabilisation and counter-narcotics policies came into being, while the counter-terrorist narrative should be seen less as a consequence of transnationalisation and more as a reaction to the negative effects of transnationalisation (in terms of the transnational dilemma) on UK policymaking and strategy. To this end, this work discusses the stabilisation and counter-narcotics policy narratives first, before moving on to an analysis of the counter-terrorism narrative.

This work employs a framework of treating the Afghan campaign as comprised of five reasonably distinct 'phases', which show how collective security dynamics affected the viability each policy over time and provided markers for changes to policy narratives in response to those dynamics. These phases are identified chronologically as occurring within (1) 2001–2003, (2) 2003–2005, (3) 2006–2009, (4) 2009–2011, and (5) 2011–2014. The first phase begins with the terrorist attacks of September 11, 2001 and ends with NATO's assumption of control over ISAF in August 2003. This phase is marked by the inception of the three policies of counter-terrorism, stabilisation and counter-narcotics and British attempts to carve out a suitable niche for its operations within the American-led multinational coalition in Afghanistan. The second phase begins with NATO's takeover of ISAF as a means of shoring up divisions within the alliance resulting from international disagreement over the legality and utility of the US-led invasion of Iraq, and ends with the preparations made by the United Kingdom in advance of its entry into Helmand province in early 2006. The third phase begins with the commencement of Herrick IV in Helmand and ends with the accession of Barack Obama to the US Presidency, signalling a change in policy orientation away from the democratisation agenda of President Bush and Prime Minister Blair to more limited aims of stabilisation and counter-terrorism. The fourth phase begins with Obama's strategic reviews to the end of reconfiguring the United States' policy aims for Afghanistan and ends with both the pared down objectives of the Lisbon Agreement in late 2010 and the assassination of Osama bin Laden in May 2011. The fifth and final phase, running from those events until the completion of ISAF 'drawdown' in late 2014, is marked by the full-scale revision of policy narratives for the campaign. Each of these phases provide context for the transnational dilemma's impact upon British policy narratives for Afghanistan and, in their latter phases, demonstrate the ways in which the British state has worked to make 'narrative' compensate for (a lack of) strategy.

The Stabilisation Narrative

The United Kingdom's narrative for the stabilisation of Afghanistan provides the core of the three narratives investigated in this work, as both the counter-narcotics and counter-terrorism narratives interweave it at various points. It is the first to be addressed in this work for two reasons, one thematic and the other practical. In the first instance, stabilisation has the longest history within British defence policy, extending back before the War on Terror in the form of the Blair Government's promulgation of an 'ethical' dimension to foreign policy and the 'doctrine of the international community' in 1997 and 1999, respectively. The following passages provide a brief overview of the evolution of New Labour's commitments to a form of 'liberal peace theory' as a means of contextualising stabilisation's role in Afghanistan. In the second instance, stabilisation is a practical choice for the first narrative as it constitutes the bulk of British and ISAF operations in Afghanistan. As such, it is also the narrative that most acutely portrays the difficulty of devising national strategy under the shaping effects of transnationalised defence policy.

This account of the evolution of the stabilisation narrative between 2001 and 2014 argues that it developed from a starting position of liberal interventionist ideology centred on the core and inalienable objective of the democratisation and development of Afghanistan, and ended as one marked by a realist predisposition, concerned with a far less expansive security-focused stabilisation agenda, defined primarily by progress made in military terms and typified by a focus on short-term counter-terrorism concerns. The British approach to Afghanistan as a fundamentally norm-driven plan can be understood in two interconnected ways. The first of these is to see the UK as a responsible and willing player within a collective security framework, following the lead of a transnationally-agreed agenda of stabilisation-based liberal interventionism. This institutional approach allows us to view the core tenets of the stabilisation approach – liberalisation and democratisation – as constitutive of alliance strategy. This is because they served the purposes of the institution itself by acting as the common denominator between NATO member states which tied them and their varied interests, values, and political sensibilities together and allowed them to act in concert. This way of looking at NATO/ISAF's Afghan strategy is one where norms have been not only the basis for the international community's relationship with the Afghan state, but have also been constitutive of the international community's ability to work together to a common goal – the stabilisation of Afghanistan – whilst also working towards often very different ends. As with any large coalition, ISAF members have different military specialisms and capabilities – ranging from counter-terrorism and military activity to provision of aid and the administration of bureaucratic reform – as well as diverse political cultures from which these activities are or are not sanctified. As a result, their activities have often bore little direct connection with one another and sometimes have conflicted. As a common denominator between these states, security via liberalisation allows for a practical 'meeting of minds' between political entities that would otherwise (and more than occasionally have) failed to agree on the ultimate purpose of intervention or the best means to go about achieving security. The liberal ideas underpinning ISAF's work in Afghanistan should, therefore, be seen as the unifying force binding together the apparatus carrying out the task.

Indeed, the centrality of liberal ideas is easily demonstrable. The democratisation and liberalisation of Afghanistan along lines amenable to the minimum standards expected by donor entities was a core proviso of most aid contributions and military commitments; every major agreement between the nascent Afghan state and the international community – from the constitutive Bonn Agreement of 2001 to the Afghan Compact of 2006 – can be viewed as 'contract-like' in their insistence on the importance – and importing – of liberal standards of governance (Suhrke 2011: 234). In other words, in very substantive ways the coalition defined the nature of the mission from the beginning in a manner that reflected the demands of maintaining a coalition – that is, by framing the mission around norms that would cohere enough participants together to make the mission possible; 'structure' defined strategy (Rynning 2012: 39–40; Ledwidge 2011: 97, Bird and Marshall 2011: 154–5). The significance of this viewpoint is not necessarily to stress the undeniable centrality of normativity in institutional behavioural dynamics, however. While it seems evident enough, that would be an unnecessary endeavour in a sense, since the allure of liberal peace theory and concomitant liberal interventionism is that it weds together rationalist and constructivist positions into a syncretic whole by claiming that interests and norms are, in fact, inseparable concepts. It posits that it is through the guiding light of norms and the spread of those norms that interests are served. What stabilisation represents is the amalgamation of two seemingly oppositional categories into a happy unity, one that promotes the individual interests all members have for security against terrorism, but in an ostensibly benign way that does not offend the sensibilities of some states in the way a coarser counter-terrorist agenda might. Thus, the alliance is secured, and intervention is possible: crucially, in this reading it is only possible because of alliance cohesion, which requires the animus of liberal peace theory.

A second way of viewing Britain's commitment to a normative-based strategy for Afghanistan relates to the idea of Britain under the Labour Governments of Tony Blair as taking a lead within the NATO alliance as a means of bridging the gap between a militaristic United States and a more development-oriented continental Europe. The idea of Britain as a 'bridge' between the two is nothing new, of course, but has increased in importance since the Thatcher Government (Gamble 2003). What is somewhat more novel is the so-called 'ethical' foreign policy of Tony Blair and his Labour Governments of the 2000s. As Jonathan Gilmore (2014: 544–5) has argued, the Labour Government's foreign policy contained within it strong advocacy of the 'human security' agenda, including 'the idea that ethical responsibilities to non-citizens deserved overt attention', and was driven by a commitment to and belief in Kantian 'universal values'. The Blair Government (or at least Tony Blair as an individual) held a set of philosophical assumptions about the transformative power of military intervention and frequently made ontological claims to the innate universality of western norms and values to all peoples. He articulated a universalist vision of western democracy where all people, given the chance, embrace freedom (Blair 2001b). The roots of this approach are located in the late Robin Cook's foundational statement on an ethical dimension to foreign policy in 1997. He argued a core element of British foreign policy – that which made it ethical – was the 'promotion of democracy' around the world, and that this was essential to Britain's national interests because the defining features of a globalised world – internationalism and interdependence – rendered the very idea of 'national' interests somewhat obsolete (Cook 1997, online). For Cook and his Labour colleagues, democracy promotion was the end of their ethical foreign policy as well as the means by which an interdependent international community must be oriented. In Blair's 1999 Chicago speech, the implications of ethical foreign policy for the structure of international coalitions and the content of liberal interventionism were made clear; a 'doctrine for the international community' was necessitated by the fact that coalitions were necessary to deal with global issues and, as a consequence, it was argued that the 'national interest [of the UK] is to a significant extent governed by international collaboration' (Blair 1999, online). Again, norms are central here: the interests of the international community were best served by respect for a rules-based system of international law, meaning that strategies of intervention could only be achieved through collective security mechanisms, which in turn would only be legitimate by basing those mechanisms upon liberal peacebuilding norms.

To this end, Labour's ethical foreign policy sought to synthesise various categories: British interests were equated with those of the international community; British values were equated with 'global values'; British strategy was a combination of interests and values, which were equitable with international interests and values. In its broadest conception, Labour foreign policy advocated the 'merger of values and interests' to the point where security and democratisation represented an inseparable unity in its own right, in which one without the other would be an impossibility: security without democracy was a false security, and democracy without security was a false democracy. By claiming that Britain's interests were also its values, and that those of this country were also those of the international community, Blair's doctrine placed Britain at the centre of, and to some extent at the mercy of alliance politics. It could lead and follow in equal measure – so equal in fact that 'leading' and 'following' ceased to be an adequate or meaningful binary. As a response to what are undoubtedly global issues requiring global responses, this approach should be seen somewhat charitably as a reasonable formula for international cooperation. The problem is one of analysis, however, since Blair's totalising worldview by design defied categorisation, asserting as outdated the strategic utility of demarcating boundaries and categories in terms of states, interests, and values. The implications of such a reconfiguration of British foreign policy should be evident when taken in relation with the statements of Inge, Stirrup and Fry quoted in the introductory chapter. It is amidst these uneasy linkages between values and interests, and national and international interests that British strategy has had to navigate to rediscover its terms of reference.

Britain's strategic position and philosophy in the 2000s has carried with it severe strategic implications that cut to the core of Britain's strategic dilemmas. First amongst these issues concerns the basics of strategy itself and whether stabilisation represents a coherent strategic concept or merely an operational approach. Strategy links means to ends: in the case of Afghanistan, the means have been primarily those of military force and development aid and expertise, and the stated end has been the stabilisation (via liberalisation and democratisation) of the Afghan state. It is not clear, however, how the use of military force can meet ends which are essentially political. Indeed, this point has been the focus of considerable academic attention in recent years, with some commentators positing the diminished 'utility of force' in an age of 'new wars' and in stabilisation operations more specifically, where the consent of the population and the linking of that consent with perceptions of state legitimacy are deemed central to success (Smith 2006). British Governments over the last decade have largely recognised the validity of the need to combine military effects with civilian socio-economic measures, and in 2005–06 offered a solution to the problems associated with the supposed diminished effect of military power by producing the 'Comprehensive Approach' (CA) framework. As the organising concept informing Task Force Helmand, the CA conceived of British military presence as a means to an end of facilitating the primary goal of reconstruction and linking improved governance to the Afghan state. The CA represented the encapsulation of the synthesis of security and development posited by advocates (but mostly critics) of liberal intervention, allowing for security concerns to be addressed via a normative framework of liberal peace.

Blair's view of the universality of liberal norms and the essential validity of liberal peace theory has been identified by several scholars as a contributing factor to the inadequacy of intelligence informing the planning process (Farrell and

Gordon 2009; Egnell 2011). It is likely that this worldview contributed to a lack of awareness or depth of appreciation for the Afghan mind set towards occupation. On the one hand this is surprising given the United Kingdom's history with Afghanistan (encompassing three previous conflicts in the nineteenth century and one proxy war in the 1980s); on the other it is not, since Labour's foreign policy outlook would undoubtedly view Britain's presence in Afghanistan in normative and legalistic terms as a liberating force there for reasons of stabilisation, not 'occupation'. Of course, this is irrelevant in terms of the development of strategy: what mattered was what Afghans thought, and the British perspective in this regard must be seen, as with Iraq, as a severe case of 'liberal optimism' (Monten 2005: 144). Intelligence failures stemming from this misjudgement fed in to incorrect assumptions regarding the size of the opposition that British forces would face upon entry into Helmand as well as the lengths to which they would go to expel the 'occupiers' (Egnell 2011: 304). The initial intensity of insurgent violence against the British was evidently not anticipated by civilian and military planners, given the focus in public statements by then-Defence Secretary John Reid that the move to Helmand could be performed 'without a shot being fired' (Reid 2006a).

Appreciating the extent of liberal normative assumptions in the UK planning process for Helmand is crucial. As with NATO, liberal norms had a constitutive and coalescing effect within the British state. As the name suggests, the CA was designed to be 'comprehensive' in that it was to link all elements of the state and their respective remits together into a coherent whole. Incorporating the work of state bureaucracies – the Armed Forces and Ministry of Defence (MOD), the Foreign and Commonwealth Office (FCO), and the Department for International Development (DfID) – with very different working cultures and priorities, the CA was a highly ambitious and perhaps unprecedented inter-departmental concept. As a doctrine for institutional cooperation centred upon the liberal assumptions of stabilisation theory, the CA viewed military force as a supporting activity and development and reconstruction as primary. In 2006, CA principles were applied to the Afghan Compact, a document of strategic intent prefacing and informing the deployment of Task Force Helmand in April of that year. The Compact was similarly ambitious in scope, containing numerous commitments to the wholesale transformation of Afghan civil society and governance. The interests of each element of the British state appear to have found their niche within this document. Liberal norms ostensibly served to unify the institutions tasked with putting them into practice. However, the coalescing of institutional interests may have worked too well: military-government relations in the lead up to the Helmand deployment have been described as one of a scenario of inadequate critique and mutual agreement between the two parties. An aide to the Labour Government during the planning period for Helmand, Matt Cavanagh, has provided an account of the process where both sides buttressed one another's optimistic assumptions about the feasibility of the mission given the resources allocated (2012: 51). Assuming the nominal veracity of such claims, the result of this environment of collective reinforcement of a institutionally self-interested way of thinking about the move into Helmand appears to have been the creation of an overly ambitious framework of complete social transformation of the province, one that was later lambasted by then-Shadow Defence Secretary Liam Fox as tantamount to applying western ideals to a '13th century' society (Fox 2009, online). This scenario would lead, similar to that of ISAF, to severe institutional conflicts of interest in the years that followed.

Thus, this chapter argues that an early narrative adherence to norms of liberal peace theory set the tone for a gradual expansion of aims and objectives within the stabilisation approach, reaching a crescendo in 2006 with the authorisation of the 'Afghan Compact', and in 2007 under Gordon Brown's prime ministership, who reconstituted and expanded the CA and promoted a vision for stabilisation that contained over a dozen different benchmarks for success. This expansionism in turn produced a state of strategic and communicative confusion over the purpose of the stabilisation mission and, taken in tandem with various political and military difficulties in Whitehall and Afghanistan, resulted in the reconfiguration of the stabilisation narrative – under the auspices of a growing institutionalisation of SC practices – to one that did away with much of the substantive content of the CA. This in turn produced a political momentum for the further narrowing of policy objectives in Afghanistan and correlated with a contraction of the stabilisation narrative to such a degree that, by 2009–2010, it no longer bore much resemblance to the original aims and rationale articulated during the Blair era, with severe implications for narrative adherence to transnational normative and institutional principles. The stabilisation chapters conclude by noting a core paradox incurred as a result of narrative revisionism: that by solving the issue of 'mission creep' created by Blair, and poor political communications as to the nature and purpose of the mission, the SC practices of the Brown and Cameron Governments created the conditions whereby the stabilisation narrative had, for all intents and purposes, negated itself.

Phase I: 2001–2003 – Balancing Counter-terrorism with Democratic Stabilisation

From the early stages of the war in Afghanistan, Blair and his Government set out a stabilisation narrative that operated from the basis of 'high mimesis' (Smith 2005) – that is, one that placed the importance of stabilising Afghanistan at the same existential level as the elimination of international terrorism. At the heart of Blair Government's approach was a Kantian discourse of democratisation and human rights as not only necessary for the stabilisation of Afghanistan but, as a corollary, as an 'antidote to terrorism' (Chin 2009: 133). Throughout his tenure as prime minister Blair frequently made reference to the importance – indeed, indispensability – of democratisation of Afghan and other illiberal societies for the project of eliminating international terrorism, often to the point where he apparently viewed this as an imperative above and beyond the short- to medium-term requirements of security and international order. In this sense we may understand Blair's philosophical motivation as stemming from a loose form of Kant's 'categorical imperative' of placing people – and democratisation of societies – as an end superior to, or at the very least equivalent to – the stability of those states and the security of the United Kingdom; to quote Kant, this imperative was that 'which declares the action for itself as objectively necessary without reference to any aim... [or] any other end' (Kant 2002: 31). Judging by Blair's predilection toward divorcing narrow self-interest from political agency (as evidenced by his declaration in Chicago in 1999 of moving the national interest in line with that of the international community), one may go forth on the premise that his view of the importance of ends as 'of themselves' was quite genuine.

It is, however, nearly impossible to speak of Blair's motivations without reference to those of the Bush Administration. Following the September 2001 attacks, the Bush and Blair governments rhetorically converged in a united front in their response to international terrorism by evoking an aggressive form of 'liberal peace theory' that viewed democratisation as 'more important than political order' (Desch 2007/08: 10–11). For Blair, democratic values were both universal – in the sense that, in his view, all people aspire to them – and particular, in the sense that the lack of such values were what distinguished terrorists from civilised peoples:

> Our beliefs are the very opposite of theirs. We believe in reason, democracy and tolerance. These beliefs are the foundation of our civilised world. They are enduring, they have served us well, and as history has shown, we have been prepared to fight, when necessary, to defend them. The fanatics should know that we hold our beliefs every bit as strongly as they hold theirs, and now is the time to show it. (Hansard 2001a)

Most importantly, Blair's normative view of the War on Terror implied that the object to be defended from al-Qaeda was not primarily infrastructure, citizens, or interests, but values. The strategic implications of a normative approach to the conflict varied between Bush and Blair, however, and as such the rhetorical similarities between Blair and the Bush Administration did not extend far into the substance of their foreign policies. While both viewed democratisation as a 'universal good' (Dunn 2003: 285), the means of achieving it differed substantially. Whereas Blair would seek to convince the Americans and the international community that 'the humanitarian coalition to help the people of Afghanistan is as vital as the military action itself' (Hansard 2001b), the Americans held what Jonathan Monten (2005: 141) has referred to as a 'vindicationist framework for democracy promotion', where 'the aggressive use of U.S. power is employed as the primary instrument of *liberal change* [my emphasis]'. Rather than implement adequate post-conflict strategies for stabilising states like Afghanistan and Iraq, the Bush Administration, populated by a coterie of neo-conservatives in the thrall of democratic peace theory, believed that democracy can be 'imposed' by military means (Dodge 2010: 1274). Again, in rhetorical terms, there is a degree of convergence between the British and American administrations here: both viewed the ultimate end of counter-terrorist operations to be a freer, more democratic world. This was not only the means to defeating terrorism but also the end of doing so. This was possible because Bush and Blair's administrations argued that democratic and humanitarian principles were innate to the human condition (Biggar 2011: 29). Since this was held to be hardwired in people, so too was the notion, according to Tarak Barkawi (2006: 56), that 'democracy and free market economics are the source of peace and prosperity for humankind' and that, therefore, 'modernizing the non-European world along liberal lines will lead to stability and development'. The significant difference between the British and American perspectives on democratisation was not so much a belief that, since such values are universal, all peoples will eventually become democratically-minded and therefore prosperous, but rather related to divergent views concerning the methods necessary to achieve that eventuality. Blair believed that reconstruction and state-building efforts would be required to set the Afghans on the correct course, and that this would be best achieved by joining the Bush Administration's traditional, physical security concerns and the coercive military apparatus with New Labour's focus on the security of liberal democratic values and a development agenda to match.

For New Labour, following up air strikes and special forces operations against the Taliban and al-Qaeda with reconstruction, development and democratisation efforts represented a near-perfect opportunity to put the party's philosophy into practice and, in so doing, showing willing to the US whilst maintaining its commitments to ethical foreign policy (Beech 2006: 116). Historical accounts of the Blair Government's relationship with the Bush Administration during the course and aftermath of the six-week invasion of Afghanistan demonstrate that this perspective was not necessarily shared by the Americans, however, whose political and military elite famously eschewed any notion of 'nation-building', and preferred instead to leave governance issues to the victorious Northern Alliance and America's European counterparts (Seldon 2005: 509–10; Hill 2005: 393–4). The Blair Government plugged this gap in US interest. In his first address to the Commons following the fall of Kabul in mid-November 2001, Blair made the case for British leadership in the reconstruction mission for Afghanistan:

> Let us be clear – the way that the world embraces and supports the new Afghanistan will be the clearest possible indication that the dreadful events of 11 September have resulted in a triumph for the international community acting together as a force for good, and in the defeat of the evil that is international terrorism. I think that we all know now that a safer world is built, ultimately, out of secure countries representing all their people living in peace with their neighbours. That is how terrorism will eventually be defeated, and that, step by step, must be the new international order that emerges from the worst terrorist outrage in our history. Whatever the challenges and whatever the setbacks along the way, I believe that is a vision and a world worth fighting for. (Hansard 2001d)

Blair made the case here for nation-building and principles of human security as the key to international security, and set the success of the mission against a liberal benchmark of representative government. He made it clear that in order to 'defeat' terrorism the international community would have to commit itself to fostering democratic governance and that, in the long run, defeating terrorism via democratisation would allow for the spread of a 'new international order'. It was from this narrative context that Blair and his Government entered into participation in the stabilisation process begun at the Bonn Conference in December 2001 and carried through to the Tokyo 'donor' conference of early 2002. The Conference centred on 'security sector reform', and was essentially a stabilisation fundraiser. This allowed US allies that shared Blair's liberal peace theory convictions to carry out the side of the War on Terror they felt comfortable participating in and, equally, freed the US from activities with which they were not comfortable. The participation of European NATO members in the state-building processes outlined in the Bonn and Tokyo Conferences therefore provided an ideological complement to the American counter-terror effort. Tokyo served two crucial institutional functions. Firstly, it connected the soft power approach of the British and Europeans with the militaristic American approach, thus strengthening the 'special relationship' between Britain and the United States whilst ostensibly tempering the latter's bellicosity and thereby strengthening the European development agenda (Clarke 2007: 604–5). Secondly, in shaping a policy response to Afghanistan defined around shared norms of democratisation and humanitarianism, it allowed a common European NATO policy towards Afghanistan to coalesce. It is doubtful whether an alliance between European states would have been possible if the mission was conceived as anything other than that of reconstruction, particularly given the various 'national caveats' of European armed forces. Indeed, the two institutional functions initially complemented one another: the Americans did not want to 'do' stabilisation, and the majority of the Europeans did not (or could not) 'do' counter-terrorism. As such, the alliance dynamic for Afghanistan was from the start a two-tier operation, comprised of the counter-terrorist, American-led Operation Enduring Freedom and the multi-national stabilisation-oriented International Security Assistance Force. The result was military cooperation based on culturally informed limits of capability: NATO members and other US allies could provide assistance on the basis of what their caveats would allow. Collective security dynamics defined the roles and responsibilities of member states, allowing each side – counter-terrorism and stabilisation – to 'free ride' on the others' capabilities.

Typically, the British took a major role in such efforts, taking on leadership responsibilities for the fledgling, UN-mandated International Security Assistance Force (ISAF) in Kabul, as well as securing 'lead nation status' for counter-narcotics responsibilities. The challenges of Blair's rhetorical approach to stabilisation became evident almost immediately during this period, however. The Bonn Conference (aimed at post-conflict resolution) was plagued with divisions within the victorious Northern Alliance, with one regional warlord, Abdul Rashid

Dostum, slamming the ensuing Agreement as unrepresentative, while former president Burhanuddin Rabbani explicitly rejected the formation of an international peace-keeping force on the grounds that it would be – ironically – destabilising to the country. Most importantly, the leading representative of the majority Pashtun population, Abdul Hadir, walked out of the Conference talks stating, similar to Dostum, that the accord was little more than victor's justice (Dejevsky 2001: 8). The lack of Pashtun representation at Bonn would prove to be highly detrimental to any hopes of political reconciliation; indeed, as one commentator presciently argued at the time, western acquiescence to a Northern Alliance-dominated post-conflict Afghanistan 'would be the coalescing of Afghanistan's majority Pashtun tribes around their Taliban leaders and the rekindling of a brutal, general civil war that would continue until the United States simply gave up' (Bearden 2001: 29). Indeed, concern for post-conflict stabilisation at this point seems to have been low, at least by the metric of monies invested: tellingly, while over $3 billion had been spent by the Americans in late 2001 on counter-terrorism operations, only $20 million had been pledged by the international community to reconstruction efforts (Sengupta 2002: 6).

Notwithstanding these early problems, following the establishment of the Afghan Interim Authority in January 2002 the stabilisation narrative was firmly entrenched and highly optimistic in presentation, so much so that British assessments of its role in ISAF hinted at completion by the end of March, with all British troops having left Kabul by mid-April (Evans and Bone 2001; Harding 2002: 20). In February, Blair sought to disassociate military effects from soft power by speaking of the impending peace-keeping and stabilisation role for the United Kingdom whilst alluding to the impending cessation of military activity. By mid-March, however, these plans were complicated by the Government's acquiescence to American requests for counter-terrorism assistance in eastern Afghanistan, and agreed to deploy an infantry battlegroup. Just as stabilisation was about to become the focus of British operations, the British presence in Afghanistan was altered to be a balance of counter-terrorism and stabilisation (Norton-Taylor 2002: 1). Oddly, whilst obviously a response to a US request, Britain was by no means forced to take on a combat role. One journalist would later report that, according to an American source, the Government were offered either an 'expanded peacekeeping role' or 'joining the combat forces' and 'opted for the latter' (Macintyre 2002: 16). Suddenly the limited military operation and long-term stabilisation plan had become an 'open-ended commitment' to counter-terrorism, according to Defence Secretary Geoff Hoon (Norton-Taylor and Borger 2002: 5). Whilst in the months previous the Government had made much of emphasising the reconstruction effort in the country, Hoon chose here to emphasise the 'war-fighting operations' the Royal Marines would be involved in (Gilligan 2002: 15). The discrepancy in Government statements about the nature of the mission led to confusion in the media as to whether the British role was in Afghanistan was primarily counter-terrorist or stabilisation in scope (Watt 2002b: 5). Compounding the confusion of Blair's comments in February of an open-ended stabilisation campaign and Hoon's comments in early March of an open-ended counter-terrorist campaign, Foreign Secretary Jack Straw claimed in a BBC interview in late March that, in fact, both operations were to be of a short duration:

> We do indeed have an exit strategy … of course we do, I don't think that our troops are going to be there, the combat troops, for very long, I, nobody can say for absolutely certain and nobody is saying that because war is uncertain business but their purpose in Afghanistan, the combat troops, is a very specific purpose to root out the remaining Al Queda terrorists and once that is done our troops will leave, and as for those troops involved in the international stabilisation and assistance force which is confined by international law at the moment to Kabul and the surrounding area, well we said we'd be there for a matter of months as a lead authority it's going to be extended for a little while but again in the long march of history this is a very limited operation. (Straw 2002, online)

The disparity of messaging in this early phase of the conflict evinces some uncertainty amongst ministers as to purpose and duration of British involvement in Afghanistan, perhaps because it was not yet clear what the British role would be beyond the cessation of counter-terrorist activity by American forces. Clearly, the scope and scale of what was required was uncertain to British policymakers, due in all likelihood and in equal parts to the volatility of the mission and the fact that strategy in Afghanistan was being largely directed from Washington. While Blair stated that he did not see any 'mismatch' with combining stabilisation and counter-terrorism operations (Watt 2002a: 12), this uncertainty was recorded amongst members of the press, who argued that Government must decide the nature of the mission it was conducting (Independent 2002: 2). The debate over the role of British forces would quickly become irrelevant, however, as the counter-terrorist mission in Afghanistan ground to a halt in mid-2002: the 'remnants' of the Taliban and al-Qaeda had failed to surface, and the key aim of American punitive operations – to capture or kill the senior leadership of al-Qaeda – had come essentially to nought. According to journalist Jason Burke (2002: 18), the dissipation of the counter-terrorist mission led US Defense Secretary Donald Rumsfeld to dictate to Coalition troops that their role was now to 'ensure that Afghanistan can develop economically and politically so the country would not become a haven for terrorists in the future'. By mid-June, the last of the British combat forces had departed the country (Carr 2002: 8). The implication here is simple: the stabilisation mission grew out of the remnants of the counter-terrorist mission and its failure to complete all of its stated objectives. For all the Blair Government's stabilisation rhetoric, the bulk of its operations in the first year of the conflict was of a counter-terrorist nature, and moreover almost entirely directed by the Americans. The mission aims and the policy narrative therefore altered to accommodate a realisation of the complexities and difficulties of attempting to 'eradicate terrorism'; until this point of epiphany, the stabilisation mission lacked serious substance. Indeed, it was not until the end of November 2002, when the United States began to seriously commit its own forces to stabilisation efforts and announced a shift in its operations to one of '75% reconstruction and humanitarian, and 25% security and combat operations' (Goldenberg 2002: 15), that stabilisation looked anything other than a rhetorical afterthought to counter-terrorism operations. It was only at this point in time that the UK, following the American lead, established its Provincial Reconstruction Team (PRT) (Norton-Taylor and MacAskill 2002: 17; Beeston 2002: 20). Once the Americans were satisfied the counter-terrorism mission was complete, Blair could return to the stabilisation narrative, declaring the need for 'a coalition to re-build the nation of Afghanistan as strong as the coalition to defeat the Taliban' (Blair 2002b, online).

UK stabilisation efforts as generally dictated by US counter-terrorism concerns is a view that can be corroborated by patterns elsewhere in the War on Terror. The role of Iraq is crucial in understanding both the strategic and narrative shift to stabilisation in Afghanistan as well as the composition of stabilisation forces within ISAF. Blair's decision to support the Americans in their diplomatic manoeuvrings against Saddam Hussein was made explicit by late September 2002, at which point he delivered his 'weapons of mass destruction' speech to the House of Commons (Hansard 2002c). Although Blair reaffirmed his Government's commitment to Afghanistan in this speech, stating that 'we will stick with them until the job of reconstruction is done', the attention of the Government was by this point firmly placed on the unfolding situation in Iraq. By early February 2003, reports began to emerge of the United States shifting its resources away from Afghanistan and toward Iraq, precipitating a visit to the Senate Foreign Relations Committee by a pensive Afghan President Hamid Karzai to plead that the Bush Administration does not 'forget Afghanistan' (Fisk 2003: 17; Reeves 2003: 15). His apprehensions appear justified in retrospect: as the invasion of Iraq was underway during March and April 2003, insurgent violence in Afghanistan rose sharply, a correlation that has been interpreted as a result of Coalition neglect (Gall 2003: 13). Perhaps reflecting the relative demotion in the eyes of the Bush Administration of Afghanistan in favour of Iraq, in May 2003 Rumsfeld stated that operations in Afghanistan had 'moved from major combat activity to a period of stability and stabilisation and reconstruction activities' (BBC 2003, online). In reframing Afghanistan as a stabilisation mission, Rumsfeld's statement can be seen in the light of diplomatic tensions between the United States and Britain, on the one side, and France and Germany, on the other, over the legality of an invasion of Iraq that led to a crisis for the future of the NATO alliance. According to Bird and Marshall (2011: 154–5), Afghanistan presented an opportunity to those European states that refused to participate in Iraq with an opportunity to redeem themselves and NATO in the process. For Blair, the potential for an enhanced focus on Afghan stabilisation allowed his Kantian moral philosophy to reassert itself and, in doing so rehabilitate the normative claims made throughout early 2003 for the invasion of Iraq. In mid-July, Blair reaffirmed his universalist, democracy-laden stabilisation narrative to the United States Congress with a forceful speech linking the war on terror with a global development effort:

> the threat comes because in another part of our globe there is shadow and darkness, where not all the world is free, where many millions suffer under brutal dictatorship, where a third of our planet lives in a poverty beyond anything even the poorest in our societies can imagine, and where a fanatical strain of religious extremism has arisen that is a mutation of the true and peaceful faith of Islam. There is a myth that though we love freedom, others don't; that our attachment to freedom is a product of our culture; that freedom, democracy, human rights, the rule of law are American values, or Western values; that Afghan women were content under the lash of the Taleban; that Saddam was somehow beloved by his people... ours are not Western values, they are the universal values of the human spirit. The spread of freedom is the best security for the free. It is our last line of defence and our first line of attack. And anywhere, anytime ordinary people are given the chance to choose, the choice is the same: freedom, not tyranny; democracy, not dictatorship; the rule of law, not the rule of the secret police. (Blair 2003, online)

Phase II: 2003–2005 – Consolidating the Rhetoric of Democratisation

Perhaps reflecting both the rapidly deteriorating security outlook in Iraq and the relative benignity of Afghanistan, there was very little focus by politicians, academics or the press on stabilisation efforts in Afghanistan for the majority of this period. In retrospect this period is of significance, however, as it was in August 2003, when the United Nations Security Council gave unanimous approval to NATO's takeover of the ISAF mission and the expansion of operations beyond Kabul, that stabilisation efforts truly began to take shape. In the aftermath of the division of NATO over Iraq, Afghanistan presented itself as a perfect storm of renewed interest from all parties. It was in this environment that the imperatives of maintaining the collective security alliance re-emerged: various non-governmental aid organisations clamoured in the press for ISAF expansion in order to extend their own operations, while the Karzai government, unhappy at international efforts and aid contributions to this point, were equally keen for the writ of its authority to be extended beyond the capital (Marsden 2003: 95). Those European states that had been unwilling to take on military roles in Iraq found reconstruction and peace-keeping efforts amenable to their own interests and political cultures, which the United States found equally useful as a means of allocating to them post-war 'mopping up' duties, thereby freeing American forces and money to concentrate on Iraq (Bird and Marshall 2011: 114–18). This in turn placated NATO as it provided it a more concrete and culturally acceptable role within the War on Terror and ostensibly solved the pan-Atlantic crisis created by the Iraq affair. Finally, the United Kingdom's political classes found solace in all these events, as it allowed the Labour Government to focus on reconstruction and development work whilst mollifying the Conservative opposition of the unity and longevity of collective security arrangements.

Thus – and again serving the interests of all involved – the political narrative emanating from Downing Street and Whitehall in the latter half of 2003 and most of 2004 was decidedly upbeat with regard to Afghanistan, and even more so by comparison with the tempestuousness of Iraq. By November 2003, Jack Straw gave a typically optimistic assessment of progress made to stabilise and democratise the country by focusing on development indices:

> Our efforts are paying off. Four million Afghan children are back at school, including girls and young women who were denied an education under the Taliban; the economy grew by an estimated 30 per cent. in 2002–03; and more than 2.5 million refugees have returned. Next month, the Afghans will decide for themselves a new constitution, to be followed by elections next year. (Hansard 2003b)

Despite such favourable portrayals by the Foreign Secretary, the United Nations Refugee Agency claimed in the same week that the security situation in Afghanistan was rapidly deteriorating, to the point where development work was fast becoming untenable (Astill 2003: 20). Meanwhile, reports of a resurgent Taliban became steadily more prevalent in the British press (Hussain 2003: 15; Burke 2003: 23). Compounding matters was the sluggish start to NATO's expansionist activities in Afghanistan, with diplomats speaking of 'limited temporary engagements' beyond the capital; a British official claimed that 'the idea [of the expansion] is to give ISAF more flexibility, not to cover the whole country' (La Guardia 2003: 14). Throughout early 2004, mixed messages continued to surface sporadically, threatening to undermine the legitimacy of the stabilisation policy narrative. On the one hand, the news was supportive: the postponement of elections in March 2004 – due to security issues related to the need to 'disarm warlords' (Fox 2004: 25) – was met just days later with announcements from the Blair Government and NATO of the expansion of ISAF in a counter-clockwise fashion around the country as a means of bolstering security to ensure the democratic process (Norton-Taylor and MacAskill 2004: 11; Beeston 2004a: 14). On the other, however, the stabilisation approach appeared to founder: the Berlin 'donor' conference in April managed to attract only £4.6 billion in funding out of a target of £15 billion (Beeston 2004b: 20). Although a mere microcosm of the stabilisation efforts as a whole, these two events demonstrated an ongoing theme of disparity between words and deeds that would come to define the conflict for its duration.

Such dissimilitude came into sharp relief in the summer of 2004 with the publication of the Foreign Affairs Select Committee report, 'Foreign Policy Aspects of the War Against Terrorism'. Following earlier comments made by Committee members in the House of Commons that Afghanistan was becoming a 'basket case' and that NATO expansion was at a state of 'virtual non-delivery', the report outlined progress made in Afghanistan as substantial but also criticised the international community for its tendency to place rhetoric above action, stating that 'fine communiqués and ringing declarations are no substitute for delivery of the forces and equipment which Afghanistan needs on the ground' (Hansard 2004c, 2004: 9). In its most critical passage, the Committee warned that '[t]here is a real danger if these resources are not provided soon that Afghanistan – a fragile state in one of the most sensitive and volatile regions of the world – could implode, with terrible consequences' (2004: 9). Blair's response to such comments was to play up achievements made whilst simultaneously playing down future expectations:

> The actual prognosis for Afghanistan is good. Sure there are big problems. These nations are failed states of total and absolute degradation. You don't turn them around in two or three years so they become first world countries en route to joining the European Union. It's not like that for a country like Afghanistan… [It is] absolutely wrong and unfortunate if people thought no progress had been made in Afghanistan over the past two or three years… There are five and a half million kids in school including over two million girls who were banned from school. The economy has grown by 30 per cent this year and is expected to grow by 20 per cent. (Brown 2004: 6)

Straw and Blair's statements utilised the same talking points: stabilisation was working, and could be evidenced by improvements in education and economic performance. However, the real test of stabilisation in Afghanistan was security, and it was partly to this end that the NATO Summit in Istanbul was convened in July 2004. Billed in the press as a 'make or break' opportunity to maintain alliance unity in the wake of festering animosities between the United States and "Old Europe" – in the main, France and Germany – over Iraq, the Summit was a nominal success in that it cemented the extension of NATO authority over the stabilisation operation in Afghanistan and secured further troop commitments from European member states (Maddox 2004: 12; Black and White 2004: 11). This amounted to the furtherance of PRT projects in the country and, importantly, additional security for the upcoming December presidential election. While the move was lauded by Blair as a major step forward in the democratising mission for the country, his former Foreign Secretary and Iraq War critic Robin Cook noted an essential paradox in increasing the international military presence in Afghanistan as undermining the stabilisation and reconstruction mission: although aid organisations required a secure environment in which to work, by linking military personnel with aid workers the security situation had the potential to deteriorate as the insurgent forces would increasingly associate aid staff and development projects with military occupation (Cook 2004: 35). A more general critical principle could also be applied here, namely that the promotion of elections as a force for legitimising ISAF and the Karzai government was not necessarily amenable to stabilisation (Barkawi 2006: 56). Of course, such an analysis rests on the assumption that the strategy was entirely dedicated to Afghanistan's needs.

The Istanbul Summit was not: it was primarily about NATO's needs and how efforts in Afghanistan could meet those needs. Democratisation and stabilisation efforts were the means to healing a fractured NATO. Supporting democratisation efforts was a way to validate Alliance members' essential unity; as such, the acknowledgement of the complexities of stabilising such a fractious state and society was secondary to the complexities of cohering a culturally differentiated NATO. Democracy promotion was the binding force for the mechanism that could deliver stabilisation to Afghanistan. It was neither the end nor the means of stabilising Afghanistan; rather, it was the means to the end of NATO unity. Thus, when Blair stated that democracy was 'the biggest blow that there can be to international terrorism', he was correct in the sense that it was a necessary normative basis for the institutional cohesion of those tasked with the job (Hansard 2004f). At the Labour Party Conference in September, Blair made his democratic peace position clearer still, arguing 'I believe democracy there means security here' (Blair 2004, online).

Given the natural charitableness of the level of expectation regarding the probity of Afghanistan's first election since the end of the Cold War, the process can be seen in retrospect as a reasonable success. The aims of security established at the Istanbul Summit were largely met, as insurgent forces largely failed to disrupt the voting process or deter a large proportion of the electorate. Indeed, it was taken as a victory for stabilisation by the international community, which in turn took this as an opportunity to begin talk of transition and of winning 'hearts and minds' in order to secure the democratic advance of the Kabul government (Meo 2005: 19). Reconstruction and democratisation became the political watchwords in the United Kingdom throughout 2005 as the Labour Government carried out negotiations with its NATO partners and senior military commanders in preparation for the deployment of British troops into the southern quadrant of Afghanistan in early 2006. Collective security negotiations dominated these proceedings, with several factors coming into play. Firstly, it was widely publicised that the Bush Administration wished to pull a substantial number of its combat forces from Afghanistan to contribute to counterinsurgency operations in Iraq, which by mid-2005 had reached a new nadir of insecurity and internecine violence (Rashid 2005: 21). Secondly, and conversely, was the alleged British military desire to get out of Basra, where its forces had been compelled to retreat from its stabilisation approach and whose presence had been rendered strategically obsolete (Harding 2005: 11). Thirdly, there was the political dimension: given its shortcomings in Iraq and the Americans' combined renewed emphasis on Iraq and loss of interest in Afghanistan (Norton-Taylor 2005: 15), the British Government felt it necessary to display leadership in the international community by taking responsibility for ISAF and leading the push to the south.

Moreover, whereas the war in Iraq had proven politically disastrous for the Labour Party and Blair personally, the mission parameters for Afghanistan were much more politically favourable in domestic terms. The aims of the ISAF expansion – reconstruction, democratisation, development and counter-narcotics –

were on paper decidedly benign, based on shared norms and fitting neatly within the confines of a Labour-friendly focus on human security. Afghanistan, unlike Iraq, remained the 'good war' in popular perception, with (at this point in time) a relatively clear and valid rationale for entry and, crucially, was untainted by allegations of illegality, ulterior and selfish motives, unilateral imperialism, and the most quintessential anti-Labour pejorative of 'spin', stemming from claims in the press that the Government's dossier outlining Iraq's weapons of mass destruction was 'sexed up'. Whilst the Butler Report that investigated these allegations did not affirm their validity, it did raise concern regarding 'the informality and circumscribed character of the Government's procedures which we saw in the context of policy-making towards Iraq risks reducing the scope for informed collective political judgement', otherwise known as Blair's tendency to favour a "sofa government" approach of assuaging concerns and offering quid pro quos (Butler et al. 2004: 10). Despite the lessons ostensibly learnt by the Blair Government as to the pitfalls of such an approach – particularly given the fact that there were indeed no weapons of mass destruction to be found in Iraq – the historical record to date appears to favour the conclusion that the planning process informing British deployment to Helmand took a similarly informal route as that of Iraq. Former Downing Street special adviser Matt Cavanagh provided an insightful account of the policy-making process, describing it as being directed by a small number of ministers, centred upon Blair and Defence Secretary John Reid, and a small number of senior military commanders. Both parties agreed on the basis of the plan, argues Cavanagh, because their respective institutional and international agendas were met in doing so:

> there was a mutually reinforcing dynamic at the top of government: almost all the key ministers thought the overall decision was clear, reinforced by their sense that the military were happy with it; for their part the senior military were indeed happy, but also reinforced by their sense that the politicians had made up their minds, so they might as well get on with it. (Cavanagh 2012: 51)

The dynamic described appears, as with the Tokyo Conference and Istanbul Summits, to be one of meeting the various institutional interests of those involved. The implications of this convergence of interests on the stabilisation narrative are stark and have persisted throughout the remainder of the campaign. In many ways it is reminiscent of the categories of 'groupthink' as expounded by Irving Janis (1972: 174–5). Reading Cavanagh's and others accounts (including former British Ambassador to Afghanistan, Sherard Cowper-Coles (2011: 11)), leads one to the not unreasonable inference that the planning process for Helmand evinced several of the core features of groupthink, including 'excessive optimism' and a 'belief in the group's inherent morality', 'collective efforts to rationalize' the intervention as being strategically viable, a 'shared illusion of unanimity' of judgement, and 'direct pressure on any member who expresses strong arguments against any of the group's stereotypes, illusions, or commitments' (Janis 1972: 174–5). The plan was based on an assumption of the correctness of liberal peace theory and democratisation-led stabilisation, and was constituted as such by a kind of institutional compromise where all parties agreed to contribute to a larger objective, but only on the basis that their specific interests would be met and maintained. Moreover, Cavanagh's account suggests that as a result of these inter-personal dynamics the strategy of undertaking reconstruction in Helmand in a 'semi-permissive environment' was evidently devised without recourse to an alternative plan; it appears to have been a result of eager politicians and civilian bureaucrats agreeing terms with equally eager military commanders.

Consequently, the tenor of the transnationally-driven stabilisation policy narrative as it developed through 2005 was one that appeared to satisfy all elements of national and collective interest groups, fulfilling both the interests of the British state's constituent parts as well as its sense of obligation within the NATO alliance. It was strategic in the same way it was comprehensive: it met the strategic objectives of those participating as much as (if not more so than) the strategic requirements of stabilising Helmand. The Government would be content that Helmand supported their overall aims of counter-narcotics and stabilisation, and would increase their credibility as a reliable partner in the eyes of the United States. The military, meanwhile, could be satisfied that Helmand offered them a chance to restore their reputation internationally and provided them with a pretext for reducing its commitments in Basra. It appears that this climate of mutual satisfaction of interests correlated with a lack of strategic consideration and, interestingly, resulted in a highly cohesive and convincing narrative with all parties agreeing on the purpose of the mission as both existential (in terms of global security) but also couched in a strong moral argument for democratisation and human rights. The reason for this is relatively straightforward: the narrative was strong because all parties agreed that it was, and the strategy was weak because this agreement via the satisfaction of various interests meant the viability of the plan was not sufficiently challenged. There were few involved in the process and those that were involved were all pushing in the same direction. The narrative therefore had a semblance of unity of effort. The strategy, however, was weak as a result of the lack of deliberation: the military goals that emerged from this unity were made to sit alongside the reconstruction aims, which were in turn assumed to be compatible with counter-narcotics objectives. This was the high point of stabilisation: the remainder of the campaign can be traced by looking at the points of convergence, departure, inflation and deflation of narrative in relation to strategy. The stabilisation policy narrative would remain unified until it became plainly evident to all concerned that the strategy it spoke of was insufficient, at which point the strategy would be strengthened but at the expense of the coherence of narrative. When it became clear that the strategy was still failing to deliver on the original rationales as set out in the stabilisation narrative, the narrative was reconfigured to meet the demands of coherence between them. The result was not only the fracturing of the CA, but of the institutional consensus that informed its creation.

Even at this point, however, there were differences of views from mid-ranking officers as to what the mission in Helmand would entail, how it could be conducted, and how long it would take. For example, in late May 2005 Colonel James Denny, the then-commander of British forces in Afghanistan, argued that ISAF's nation-building responsibilities would require 'a generation' to complete before withdrawal would be possible, whereas the projections given by politicians was a mere three to five years; another British officer within NATO, Colonel Huw Lawford, commented that the mission would not be benign, stating that '[y]ou will not be going out in Land Rovers, you will be going out in armed Warrior vehicles, and you will not be walking around in a beret, you will be going out in a tin hat, with a rifle and body armour' (Bentham 2005: 21). This plainly jars with estimations made by Reid and others at the time, who spoke of an aim of 'building a democratic, pluralistic, and politically and commercially non-corrupt Afghanistan' through a strategy of 'build[ing] up the economy, civil society and security forces of the Afghan Government' (Hansard 2005c). Those at the upper echelons of the state spoke the language of a relatively benign, post-conflict environment generally conducive to stabilisation, whilst those at the operational level appear to have viewed the deployment as involving a greater degree of kinetic effort. It would seem that the Blair Government was either unaware of the nature of the mission confronting them or, as Betz and Cormack (2009: 326–7) argued, that their reconstruction rhetoric evinced 'severe strategic lassitude' in that they 'wanted to play the reconstruction theme of the opera and did not overly concern themselves with how or by whom the other elements of the orchestra would be conducted'. Interestingly, Betz and Cormack's view intimates that the demands of a strong narrative undermined the development of a strong strategy. In this reading, ministers' views on the deteriorating security situation on the ground appeared entirely out of step with those of the military commanders in-theatre: Reid spoke of his belief that British forces would be given 'a great welcome' by Afghans, while Minister of State for the MoD Adam Ingram argued rather counter-intuitively that the stabilisation strategy was appropriate because the placing of forces on the ground to facilitate reconstruction would cause the conditions required for reconstruction to worsen: he did 'not believe the situation is deteriorating. Incidents can occur because contact has been made with people who are being pursued because they are intent on making the situation worse' (Hansard 2005a). For the remaining four and a half years of Labour government, the sense of Government as out of step with reality (or of trying to mould reality to fit the preferred narrative) would become ever more pronounced, stretching civil-military relations to near breaking point and, as a consequence, producing the circumstances that would produce an incoherent stabilisation policy narrative.

Chapter 4

The Fall of the Stabilisation Narrative

In this second stabilisation chapter, I chart the demise of the high idealism of the UK's stabilisation narrative for Afghanistan. Upon the entry of British forces into Helmand Province in April 2006, the assumptions underpinning the democratisation and reconstruction elements of Labour's stabilisation policy came into contact with the harsh realities of southern Afghanistan for the first time. This chapter argues that the weight of events in Helmand rapidly undermined the Blair Government's 'Comprehensive Approach' and resulted in a gradual yet unmistakeable reduction of stabilisation aims in the narrative, ultimately resulting in a revised, national security-centric narrative from late 2008 onwards. This trend may be understood as the result of the implementation of SC processes as a means of overcoming substantial intra-state tensions regarding the purpose of the Helmand mission, which were in turn the result of Government attempting to meet its transnational obligations to the United States and NATO.

Phase III: 2006–2009 – Stabilisation and the Breakdown of Civil-Military Relations

The months preceding the arrival of Herrick IV to Helmand province in April 2006 were fraught with diplomatic manoeuvrings as the Blair Government struggled to create a substantial NATO coalition for ISAF's expansion to the south of Afghanistan. With the French, Germans and Spanish refusing to contribute in a war-fighting capacity, and the Dutch government being forced to put the deployment of its armed forces to a parliamentary vote – which eventually passed – the British had to rely on a force primarily of smaller states such as Denmark, eastern European NATO newcomers like Estonia, and other Anglophone states, namely Canada and Australia (Norton-Taylor 2005: 21; Castle 2006: 2). Despite the by-then common problem of alliance cohesion, the issue of deployment to Helmand went largely uncontested in the United Kingdom; aside from a handful of critical opinion pieces in *The Guardian*, countervailing voices were largely absent and the press were largely supportive of the mission (Independent 2006: 26; Daily Telegraph 2006: 23). Similarly, despite their future position of criticising Labour's 'dreamy ideas' of stabilisation, the Conservatives were amply collaborative with Government upon the announcement of the mission by John Reid to the Commons, with shadow Defence Secretary Liam Fox opining that the 'defence of our national security and the construction of a free and democratic Afghanistan are noble ideas shared on both sides of the House' (*The Guardian* 2006).

The London Conference in late January coalesced the ISAF forces moving south around an ambitious programme of state-building that represented nothing less than a commitment to the wholesale transformation of Afghanistan's political, economic and social structures. With most targets set for the end of 2010, the ensuing 'Afghan Compact' bound the Afghan government and the international community together and reaffirmed in the British Government the centrality of liberal normativity and adherence to collective security principles to the stabilisation policy narrative, and contained a raft of measures aimed at such a policy, including establishing a functioning judicial system, curbing the narcotics industry, 'ensuring macroeconomic stability', reducing corruption within the state, developing the country's 'human, social and physical capital' and promoting 'democratic governance and the protection of human rights' (London 2006: 2–5). In early February, Reid addressed the Commons with a finalised statement that described British troops' role in Helmand in light of the Compact as one of 'working to ensure that we provide Afghanistan with a seamless package of democratic, political, developmental and military assistance in Helmand' (Hansard 2006a).

In mid-February – just a fortnight after the Compact was agreed upon – relations between the military chiefs and government ministers took a negative turn as Blair apparently reneged on the alleged civilian end of the Helmand bargain by refusing to withdraw troops from Iraq until the security situation there had improved (Rayment 2006: 1). This led to a series of newspaper articles exposing the mixed messaging between the Armed Forces and the Ministry of Defence: while senior military figures including the former Chief of the Defence Staff, Lord Guthrie, spoke out publicly warning of 'overstretch' from fighting wars on two fronts, Reid argued the defence chiefs had informed him that the Helmand operation was manageable without withdrawing forces from Iraq (Sengupta and Taylor 2006: 19; Rayment 2006: 1; Norton-Taylor 2006: 11). It is perhaps telling that in the light of public rebukes that the Government had neglected its duty of care by providing inadequate resources to military operations, the period of late February and early March featured Reid making more frequent statements reiterating the peace-keeping aspect of the mission, most famously articulated in his misconstrued quotation highlighting his desire to 'leave in three years without a shot being fired':

> The mission is quite clear. We'd go there not to hunt terrorist[s], though if we're attacked by terrorists or insurgents we will obviously defend ourselves. We go there to defend President Karzai's government, the democratically-elected government of Afghanistan, and civilian authorities who are helping him from the whole international community, to build their economy, build their democracy, and build their own security forces. (Reid 2006a, online)

Here Reid makes clear the difference between counter-terrorism and stabilisation. At the same time, however, it is apparent that although Reid did frequently warn of the potential for attacks against British personnel, there was little in the way of contingency planning that could alter the mission accordingly if it was deemed that a reconstruction approach was no longer suitable. Almost immediately upon commencing operations under Herrick IV, however, Reid's stabilisation narrative was proven irrelevant as British forces found themselves engaged in fierce battles across Helmand province and focused on force protection rather than on reconstruction efforts. Senior officers and infantrymen alike expressed their concerns to the press that the mission had lost its purpose practically on arrival: one stated that he believed the 'government is hiding the truth from the public' to save Blair from criticism, while another lamented that the government needed 'to decide what our mission out here is – because we can't do hearts and minds and this (fighting). It just won't work' (Smith 2006: 4; Walsh 2006: 18). The appearance of statements such as this in the press represented a fundamental challenge to the Government's stabilisation policy narrative. Initially, however, they appear to have had little effect: the official response to the rapid (and public) unravelling of the Taskforce Helmand strategy was to deflect attention from the perceived inadequacies of the planning stage and to reinforce the reconstruction narrative in spite of the fact that, at that time, intense kinetic activity had precluded almost all hope of reconstruction. While it is true that the reconstruction element of stabilisation must logically be preceded by fighting to pacify the area in question (as the 'clear, hold, build' approach makes clear), the intensity of the fighting taking place and the almost total absence of reconstruction work during the early stages of the Helmand mission appears to have undermined perceptions that the plan was workable. This did not give cause within Government to change its rhetorical approach, however. Altering strategy – or at least the presentation of it – was apparently being sacrificed in order to preserve narrative, despite ample evidence showing the latter to be intrinsically untenable.

Criticism of the stabilisation strategy came from outside the military as well. Michael J. Williams (2011: 72–3) described how, within weeks of the start of Helmand operations, the 'Stabilisation Unit' (the successor organisation to the PCRU) reported that 'UK objectives in Helmand [were] not achievable' as formulated in the Afghan National Development Strategy (ANDS). This implied that not only was the Helmand plan untenable, but by extrapolation that the liberal normative basis for it was flawed. It is probable that, even if Government had committed the entire resources of its Armed Forces to Afghanistan from the beginning, it would have still been insufficient to stabilise the province – the population of which was well in excess of one million persons – in the manner called for by the ANDS. Moreover, the plans made for stabilisation were inherently state-centric – specifically, focused on extending Kabul's authority to Helmand – a concept of governance essentially at odds with the social mechanics and cultural sensibilities of the province. In spite of this advice – or perhaps because of its catastrophic implications – Cabinet appears to have simply ignored it, preferring to continue to speak the language of reconstruction irrespective of the feasibility of policy. As though to downplay the significance of planning altogether, the newly appointed Defence Secretary Des Browne would push the bounds of credulity by commenting in July that the 'deployment was always going to inform us better than the pre-plan part of the assessment' (Wintour 2006: 18). Thus, suspicions that the strategy was 'ad-hoc' were countered by Browne by claiming that 'ad-hocery' was a natural part of the planning process. Perhaps most perniciously of all, Browne would also claim that the mere questioning of the Government's pre-war planning acumen would be tantamount to endangering the lives of British military personnel:

> If there are suggestions of confusion, or … that we are there primarily to do something entirely different, that is played back by the Taliban into their communities and people think these British soldiers are coming to starve them or attack them, then that is putting our soldiers at a level of risk that is unnecessary. (Wintour 2006: 18)

To translate, the ostensible logic of Browne's statement was that the stabilisation policy narrative must be protected at all costs, and that criticism of the Government's planning process was tantamount to treasonous behaviour. It was in this politically charged atmosphere that, throughout the summer and autumn of 2006, public animosity between military officers and government ministers continued to grow. The former made more frequent comments to the effect that the government was inadequately funding military operations, and that it was without strategic direction: retired Colonel Tim Collins would comment to the press that 'the government has no idea of what it wants to do. It's invited the Army to go to Iraq, to Afghanistan, and do stuff. It would be a bit like giving your keys to builders and say go and do some stuff in my house' (Norton-Taylor and Vasagar 2006: 16). Despite such warnings from officers about the untenable nature of reconstruction as then conceived, ministers continued to talk the language of a conflict defined not by battle but by social development, once more in ways that represented their own institutional interests. For example, International Development Secretary Hillary Benn would claim in November that 'the most significant change since the fall of the Taliban is that girls can go back to school again' (Benn 2006, online). Blair's optimism remained resolute, and he would issue a statement at the end of NATO's November Riga Summit claiming that 'I think there is a sense that this mission in Afghanistan is not yet won, but it is winnable and, indeed, we are winning' (Sengupta and Castle 2006: 36). Taken in tandem, the Government's approach throughout 2006 appears to have prioritised the coherence of political messaging about the campaign above the changing dynamics at the operational level. While military commanders were compelled to publicly state their fears over a lack of strategic direction, the Government sought to downplay any assertion that the stabilisation narrative they had employed since 2002, grounded as it was in democratisation, development and humanitarianism, was no longer fit for purpose. For Blair in particular, liberal peace theory remained central, even in the face of seemingly insurmountable countervailing realities. By early 2007, responding to increasing calls to abandon liberal interventionism, he expressed his exasperation by remarking that he found it 'so difficult when people say, the situation in Iraq or Afghanistan is really challenging and difficult, therefore we should remove ourselves. Surely, the first question to ask is, do the people in those countries want democracy and freedom, answer yes they do' (Blair 2007, online). Blair's commitment to liberal norms precluded him from stating that security should be the first question, and his narrative line of democracy and freedom for Afghanistan as the end of British operations in Afghanistan remained undiminished.

Upon Blair's departure from Downing Street in June 2007, it had become apparent that the strategic approach to Afghanistan needed to be upgraded from mere rhetorical gesture. His successor, Gordon Brown, would spend the remainder of 2007 undertaking a policy review process that, contra Blair, assumed that the people of Afghanistan's first priority may be grounded in more practical issues like security than those more abstract issues relating to the merits of liberal democracy. Indeed, a review was necessary for reasons other than those of narrative presentation. Brown inherited Helmand at a time of transition: in February Britain had handed command of ISAF over to an unimpressed United States after the embarrassing loss of Musa Qala to insurgent forces and, perhaps in part as a consequence, operations in Helmand had taken on a far more aggressive counterinsurgent tactical approach (Sengupta 2007: 2). Partly due to troop increases and a concomitant rise in contact with insurgents, troop casualties across the summer months of 2007 had risen by more than 50 per cent compared with 2006, and, indicatively, the amount of bullets expended by British troops had increased sharply under the first Herrick mission under Brown's leadership (Herrick VI – 12 Mechanised Brigade) (Bird and Marshall 2011: 176–7; Fergusson 2008: 324). Despite the fact that British forces were now defeating the Taliban in almost every military encounter, brigade commander John Lorimer would famously refer to the fighting process as 'mowing the grass', in a reference to the difficulties encountered in implementing the 'clear-hold-build' approach central to the CA, under what was widely perceived as conditions of insufficient force-to-population ratios (Wright 2009: 21; Farrell and Gordon 2009: 22; King 2010b: 317). By the time Brown came to office, then, the realities of the conflict and the practical limitations of the CA had begun to impress upon the minds of leaders of the international community. Just a week after his premiership began, an UN-led conference on reform of the Afghan judicial system took place in Rome, devoid of much of the optimism of London 18 months previous. A major issue at the conference was the impact of rising Afghan civilian casualties on the stabilisation effort, with UN Secretary-General Ban Ki-Moon voicing particular concern (Ban 2007). Perhaps as a result of the uneven toning down of Blair's totalising normative rhetoric, messages from the Brown Ministry appeared mixed following Blair's departure. Some, such as Des Browne, had noticeably toned down the Kantian content of their public statements in favour of a more measured approach, and spoke less of transforming Afghan society and instead focused on helping the Afghans 'have the best future they can have in an already challenging environment' (*The Guardian* 2007). Others, however, such as Foreign Office minister Kim Howells, remained trenchant of the view that Afghanistan should become 'an independent democratic state' (Howells 2007: 43). Whereas departmental representatives pressed the relevance of their respective priorities to the UK Afghan strategy, Brown himself was conspicuous in his lack of policy statements on Afghanistan for the first five months of his tenure. Meanwhile, the press had begun to call for a shift in Afghan policy toward 'an approach based on realism rather than blind optimism' (*Independent* 2007). The problem for the United Kingdom was that 'blind optimism' was not just a rhetorical issue, it was the foundation of British strategy in Helmand, insofar as strategy required rhetorical adherence to the philosophical obligations binding together the collective security mechanisms on which the national interest depended.

In early December, following a state visit to Afghanistan, Brown addressed the Commons with a definitive statement on his government's policy for the conflict. Taking place on the same day as the re-capturing of Musa Qala from insurgent forces, Brown's speech outlined his 'comprehensive approach' to Afghanistan and, in a manner typical of his attention to detail, was delivered meticulously, focusing on a raft of areas deemed necessary to stabilise the country. These went beyond military prosecution of the insurgency to include counter-narcotics, promotion of good governance, training the nascent Afghan National Security Forces, regional engagement with Pakistan and Iran, the alleviation of poverty, reconciliation with the rank-and-file of the insurgency, and a raft of long- and short-term development and aid projects designed to address the day-to-day concerns of Afghan civilians (Hansard 2007d). Crucially, Brown spoke of his government's approach as one of 'hard-headed realists, not idealists', effectively signalling a move away from the grand rhetorical gestures of the Blair era (Hansard 2007d). For all the differences between Brown and Blair in terms of narrative emphasis and style, however, the former could not easily escape the political inheritance of his predecessor: the speech remained firmly within the confines of a worldview that posited the centrality of a democratic Afghan state and adherence to its liberal constitution as key to defeating the insurgency and eliminating terrorism, as well as the means of maintaining the legitimacy of both the British mission and the Karzai government. Nonetheless, Brown's statement garnered encouraging words from the Opposition leader David Cameron, who noted his satisfaction that Brown had apparently moved away from the Blairite devotion to imposing 'a fully-fledged western democracy in a deeply traditional society' (Hansard 2007e).

While Brown's policy refresh possibly served as an improvement over the strategically shallow orating of the Blair Government, his detailed articulations of what was required for success in Afghanistan had its own drawbacks. To recall an earlier point, the strength of Blair's argument was precisely that: *the argument*, focused single-mindedly on a narrative of democratic peace as a panacea to the ills of post-modernity, upon which was built the normative basis of collective security, and through which Britain's military activity could be morally justified. The strength of this argument was, therefore, not simply its communicative power to domestic audiences: it was also to be found in its ability to define and defend British interests as consistent with those of NATO and the

US. Brown's 'comprehensive approach', whilst of almost infinitely more value in terms of the substance of strategic and operational planning, was devoid of the style and panache associated with Blair's rhetoric. He did not forcefully make the case for the UK's role as an international collaborator in the same way as Blair had. Explaining the importance of a collective security operation premised on foreign internal defence, where the national interest is defined by another nation's security, stability and democratic credentials, and which could only be secured by panoply of stabilisation efforts carried out in conjunction with a myriad of state and non-state collaborators, is a confusing and imprecise venture. By contrast, explaining a conflict as an existential, Manichean struggle between the forces of good and evil within a transnational framework bound together by commonly held democratic values, as was Blair's (and Bush's) wont, was far more amenable to the public imagination, despite its almost complete lack of utility to the realities of the conflict.

Brown's focus on the details of the strategic and operational requirements for victory in Afghanistan came at the cost of a lack of effective communication: it was neither grounded in the cultural coding of Blair's liberal internationalism, nor could it be explained simply or with reference, in the first instance, of why it mattered to British interests. Instead, it drew its conclusions of its significance by degrees: Britain's national interest required participation in collective security operations, which thereby secure Britain itself, and in doing so it participates in Afghanistan, with the aim of democratising and stabilising the Afghan state, so that it might provide the economic, social and institutional reconstruction efforts necessary to pacify its people and tangentially defeat an insurgent force which, by virtue of its previous association with the terrorist group al-Qaeda – which attacked Britain's chief ally and guarantor of its security – poses an indirect threat to the United Kingdom by its continued presence in Afghanistan where, if it is not defeated or sufficiently marginalised, may in some hypothetical future scenario allow the return of said terrorists, which would then destabilise the Afghan government and potentially create an 'ungoverned space' from which to plan and carry out further terrorist atrocities, which may be targeted at the United Kingdom. Blair, by contrast, summed up his policy in far fewer words: 'democracy there means security here'. Brown's decision to continue with the strategic foundations for stabilising Afghanistan as established by Blair, but also to abandon much of the rhetorical basis for the accompanying policy narrative (coupled with a rapidly deteriorating security situation and a corresponding sharp increase in British casualties) meant that he would struggle to conceptualise an effective communications approach to the conflict.

As such, Brown's narrative approach was essentially honest yet unfathomably labyrinthine. Much of his prime ministership should be seen in the light of an equally uncoordinated approach to government communications, where each relevant member of his Cabinet would focus on their own compartmentalised message. His focus on the minutiae of stabilisation can be seen as giving succour to the already established patterns of institutional self-regard. Ultimately, this is the historical milieu from which the institutional practice of SC emerged, as an attempt to synchronise political communications to avoid 'information fratricide'. For most of 2008, however, this had yet to take effect and, no doubt because the Government's comprehensive approach placed a premium on highlighting the non-security, developmental aspects of Britain's endeavours in Helmand, the stabilisation narrative took on a rather shapeless form. The triumvirate of ministers delegated by Brown to lead on the comprehensive approach – Browne, Foreign Secretary David Miliband, and International Development Secretary Douglas Alexander – all continued the trend of displaying a penchant (quite naturally given the circumstances of Brown's policy) for speaking about the developmental indicators of progress even while the insurgency was escalating.

For example, in a February 2008 interview, Alexander was presented with a point of view by his interviewer that, given the deterioration in security, 'things are going pretty badly out there'. The minister responded by saying 'I would actually disagree with that description' before going on to base his argument for 'real progress on the ground' on the increased uptake of girls in school (Alexander 2008, online). Education was important to Alexander because it was a DfID issue. Miliband, meanwhile, spent much of late 2007 and early 2008 on diplomatic speaking tour duties, outlining the need for a political process centred upon effective delivery of the comprehensive approach. Curiously, despite Brown's Commons attempt to clamp down on 'idealism' within his Government, Miliband spoke frequently on this tour of his party's 'universal message'. Whilst making the case for delinking the Government's democratisation language from that of Blair and the neoconservatism of the Bush administration, Miliband nonetheless made a point to argue that 'human dignity and human rights, democratic accountability and checks on arbitrary power' were 'universal values' that were 'real and popular', and that it was therefore 'not "western" to assert' them in foreign policy (Miliband 2007b, 2007c, 2008a, 2008b, 2008c, 2008d, 2008e). Again, this reflected the minister's institutional portfolio: the promotion of liberal democratic governance was an FCO area of responsibility. In Miliband's defence, however, his rhetoric was significantly weightier than Blair's when regarding the specifics of implementing his version of liberal peace theory, arguing that 'democracy which lasts requires both a state that has the capacity to offer protection, welfare and justice to its citizens, and the accountability to ensure it acts as a servant, not a master, of the people' (Miliband 2008c). In a sense, Miliband's rhetorical positioning was symptomatic of the Brown premiership in general: big on specifics and practicalities, but still ideologically and historically wedded to the grandiose promulgations of the Blair era.

Over the spring and summer of 2008, however, the compartmentalisation of messaging according to institutional interests and a general narrative emphasis on the finer points of stabilisation theory would begin to dissipate in the face of mounting criticism of the Brown Government's handling of defence matters by the broadsheet press and the Conservative opposition. Britain's collective security obligations grew increasingly onerous over this period and consequently sowed even more discord in government-military relations. Britain's decision to begin the process of 'overwatch' in Iraq in 2007 meant that by mid-2008 there were still 4,000 troops in Basra, representing a much higher level than had been desired or anticipated by military commanders during the planning stages for Helmand (Evans 2008a: 22). This was compounded by the failure of April's NATO summit in Bucharest to secure substantial resolution to the ever-present issue of national caveats and an increasing sense that Britain was bearing a disproportionate burden of the combat in Afghanistan (Kirkup 2008a: 20). In the wake of these two events, Brown faced a series of challenges to his reputation from the press that he was sacrificing Britain's (specifically the military's) interests to a collective security framework characterised by European 'free riding', and from the senior military in his inability to match his acquiescence to collective security demands with adequate military resources and equipment. The obligation-cooperation cycle had ensnared him in a double bind reflective of the transnational dilemma: the necessity of contributing versus the political and financial costs of operations.

The pressures of transnational policy were duly exploited by Brown's political enemies. This new round of recrimination began with Shadow Defence Secretary Liam Fox's public insinuation that, in apparently advocating closer European Union military integration, the Prime Minister was undermining NATO unity (Fox 2008: 25). This was soon followed by then-Chief of the General Staff Richard Dannatt commenting to the press that the Government was underpaying soldiers by noting that they earn less than traffic wardens; Fox added to the chorus of criticism by stating that such facts demonstrated 'how badly the Military Covenant has been broken' by the Labour government (Kirkup 2008: 6). Brown's perceived failure to rectify the excesses of policy and insufficiencies of resourcing that plagued the Blair era, and his decision to block Dannatt from becoming Chief of the Defence Staff in retaliation for his remarks to the press (Smith 2008: 7), appear to have produced something of a media pincer movement against his leadership for the remainder of 2008, with high-profile former military commanders including Dannatt, Lord Boyce, and General Mike Jackson, the senior Conservative leadership, and the centre-right broadsheet press coalescing to form an ad-hoc coalition (Coughlin 2008a: 22; Evans 2008b: 21). Amidst recriminations of undervaluing the Armed Forces and short-changing the defence budget, matters were compounded by the emergence of the September 2008 financial crisis, which further tightened the Treasury's purse strings against increased defence spending (Cornish and Dorman 2009: 247–8). As well as damaging the reputation of Government in its handling of Afghanistan and supposed mistreatment of the Armed Forces, this very public falling out between Downing Street and military brass would have profound consequences at the level of strategic planning and coordination; according to Dannatt, he and Brown did not speak to one another for six months over late 2008 and early 2009 (2011: 433). The pressures incumbent upon Downing Street to meet the demands of collective security membership whilst satisfying the institutional interests of the constituent parts of the UK state were now obvious. British leadership on Afghanistan was not merely strained; it was essentially bifurcated between its military and civilian components.

Against this backdrop of disunity and narrative fragmentation, Des Browne departed as Defence Secretary in October and was replaced by John Hutton, signalling a step change in Government relations with the Armed Forces and, crucially, in terms of the stabilisation narrative, between the Ministry of Defence on the one hand, and the Foreign Office and the Department for International Development on the other. Hutton's view was markedly more sympathetic to the military's perspective than that of Brown and Browne (understandably, perhaps, given the fractious relationship that had escalated hitherto) and his policy articulations appear more conciliatory to Army interests for a reframing of the stabilisation mission along lines more amenable to military prosecution. The substance of ministerial debate between Miliband, Alexander, Hutton and Brown, and the policy implications of Hutton's arrival is imprecise, but the effect on the trajectory of the stabilisation policy narrative was stark. Within days of his appointment as Defence Secretary, Hutton would recalibrate the messaging of the mission to one that emphasised, 'first and foremost', 'the UK's long term national security interests' (Hutton 2008a, online). Hutton shifted the locus of intervention away from what it meant for Afghanistan to how it served British counter-terrorism requirements. His apparent reversion to prioritising military effects in the stabilisation narrative seems to reflect a growing mood within Whitehall that the comprehensive approach had begun to spiral out of control.

The crucial turning point in this process of narrative revivification occurred on Armistice Day, when Hutton delivered a keynote speech to the International Institute for Strategic Studies (Hutton 2008b, online). This speech reoriented the defence approach to Afghanistan by framing it unreservedly as a matter of 'vital' national security rather than as a democratisation mission or predominantly in terms of a collective security operation. This can be understood as a consequence of what he saw as the problem of 'free riding' in NATO. A narrative divestment of UK contributions in terms of the language of collective security is understandable in the context of a situation where others were not playing their part: articulating Afghanistan in terms of self-interest made more sense than explaining it in terms of collective sacrifice when Britain was sacrificing more than most of its partners. In the face of multiple failed attempts since 2004 to impel partner nations to improve their capabilities, it is reasonable to posit that a distinctively 'national' approach was a tacit acceptance of this unresolved issue (Prince 2008: 18). The pronounced use of 'security' is also interesting in the specific historical context: while Hutton desired to increase resources to the Armed Forces, he would announce significant budgetary cuts to defence projects just one month after the IISS speech (2008: 18). Given such circumstances, the emphasis on military means appears something of a *quid pro quo* with his military colleagues: if the resources for a comprehensive approach would not be forthcoming, then the narrative focus for stabilisation should be narrowed to what could be reasonably achieved. Indeed, an Afghan strategy dependent upon impending machinations in the American presidential race implies there was little strategic work Britain could do other than tinkering with narrative.

This kind of work has significant utility however: reframing the justificatory framework for Afghan stabilisation as fundamentally about British interests – rather than being about collective or Afghan concerns – contributed to the stabilisation of Government as well. Firstly, it went some distance in repairing civil-military relations insofar as it created the narrative impetus for circumscribing the aims of the mission to those the military could realistically achieve. By framing the mission in 'national security' terms and by highlighting the military aspects of the mission above the 'soft power' elements, the Army would be partially placated and the chance of another 'Dannatt' – a military officer going 'rogue' in narrative terms – was significantly reduced. Secondly, and as a result of the narrowing of discourse, it would nominally pacify all parties by producing a narrative pathway that the public could support, by arguing that the losses incurred in Afghanistan were necessary for the safety and security of British citizens. This would benefit the Prime Minister, who was seen by several in the Armed Forces as uninterested in Afghanistan but deeply concerned about its effect on public opinion. Thirdly, and perhaps most significantly, Hutton's narrative reconfiguration allowed the state to sidestep the transnational dilemma by affording ultimate ownership of the stabilisation narrative to the Army and Ministry of Defence: by placing a national security emphasis on the narrative, the burdens of collective security membership could be alleviated by keeping true to international commitments whilst denying their fundamental shaping power on operations.

Hutton's repositioning of the stabilisation narrative appeared to strike a chord with the centre-right broadsheet press, with one journalist commending him for his 'plain-speaking' amidst a Government suffering from 'intellectual confusion' about the purpose of the mission, and castigating those within the Department for International Development for continuing to frame Afghanistan as a reconstruction mission (Coughlin 2008c: 26). Notwithstanding the revolutionary significance of Hutton's narrative approach within the confines of the Labour government, however, the historical record demonstrates that recourse to a narrative of national security and military prioritisation was, in fact, a creation of the Conservative Party. Indeed, while Hutton coined the unusual term 'national security interest', it was in June 2008 that Liam Fox first articulated the national security approach, a full four months prior to Hutton, stating that 'we are in Afghanistan for reasons of national security, to deny a safe haven to those who would commit indiscriminate acts of terrorist murder on men, women and children' (Hansard 2008a). What Hutton's reframing of narrative really signified, therefore, was the outmanoeuvring of the Labour Government by an *ad hoc* coalition of Conservatives, generals, and the centre-right broadsheet press, and the completion of a pincer movement that undermined the reconstruction and development aspect of the stabilisation narrative and replaced it with a realist national security rhetoric. The demands of adhering to NATO defence policy had created the condition for treating the satisfaction of US expectations as paramount, which produced the opportunity for the perception of a unique stabilisation role for the UK that was actualised by a cohering of institutional interests to that end. When this approach began to fail on the ground, as it almost certainly would since it had institutional interests and not Afghan interests at the forefront, the institutions whose interests were seemingly forsaken by Government sought revenge. Although not entirely apparent at the time, the implication of this fracturing was broader still, signalling the beginning of the end for the entire normative element within liberal interventionist discourse. As a result, the stabilisation narrative would soon be completely turned on its head.

Phase IV: 2009–2011 – National Interest and 'Managing Expectations': Strategic Communication Arrives

In this penultimate phase of the Afghan campaign, bookended by Barack Obama's ascendance to the American presidency in January 2009 and the killing of Osama bin Laden in Pakistan in May 2011, the British narrative for the stabilisation campaign in Afghanistan completed its metamorphosis. It began as one that embraced collective security by emphasising the democratisation of Afghanistan as an end in itself and as a key foundation of global security, and ended as one that sought to traverse the transnational dilemmas of expectation, obligation, cooperation and free-riding by prioritising only the stability of the country as an end of British national security interests. The narrative proceeded from one that was grounded in internationalism and liberal peace theory to one based almost solely on principles of *realpolitik* and rationalist national security concerns. Fundamental to this process of narrative transformation was the emergence of SC as an institutional force within Whitehall, and the Ministry of Defence in particular, as the British sought to come to terms with the strategic changes wrought by the incoming Obama administration and the shifts in narrative emphasis that would ensue.

As with all other phases of the conflict, the role of Iraq was pivotal in determining the strategic course of events in Afghanistan during this period, and as a result much of the Obama administration's actions in 2009 can be seen as an attempt to transplant lessons learnt in the former warzone into the latter. On an operational level this meant attempting to replicate the 'Surge' process, but on the level of political and 'strategic' communication this translated to an attempt to 'lower expectations' regarding what could be achieved in Afghanistan. According to Thomas Ricks (2009: 165), the so-called 'COIN revolution' within the American military establishment (populated with ranks of 'soldier-scholars') brought a more nuanced, culture-centric approach to counterinsurgency operations that stood in sharp contrast to the rather monolithic worldview of the neoconservative old guard within the Bush Administration (Lindsay 2011: 769–73). Ricks and Barkawi (2006: 56) have both noted that these scholars eschewed the notion that democratisation and stabilisation were naturally supportive concepts; in fact, as Steven Metz has argued, '[f]ew things are more destabilizing and prone to chaos than democratization' (Ricks 2009: 165). This represented a major change in policy, which in both Iraq and Afghanistan had to this point remained one rooted in ethnocentric assumptions about the universality of Western cultural and political norms; as one scholar noted, '[t]he tendency of NATO has been to make assumptions about what it is that Helmandis want, as viewed through a westernised democratic lens, and to apply these assumptions to all and sundry, when the reality shows that no such template can or should be imposed' (Holland 2008: 48). From an institutionally-oriented perspective, this is only half the story: viewing what Helmandis want through western lenses should be seen as equivalent to simply pursuing what Western states wanted. Judging by the content of Obama's first major Afghan policy address in March 2009, these principles were indeed to be taken forward in the stabilisation campaign; there were no mentions of the necessity of 'democracy' to the stability of Afghanistan to be found (Obama 2009, online). Indeed, the understanding garnered from Washington in the weeks preceding Obama's announcement presaged this change of approach towards a 'new realism' where democratic movements would be supported but not directly imposed (Traynor 2009: 18; Patterson 2012: 33; Etzioni 2012: 86).

What this equated to in narrative terms was akin to a genre shift in Washington from the high mimesis of democracy as the panacea of violence and instability to a low mimetic framework of national security and stabilisation as a regrettable necessity. As if to underline this genre shift, just over a week after Obama's inauguration his incumbent Defence Secretary Robert Gates provided a clear indication of the future direction of American foreign policy in Afghanistan and elsewhere by drawing a line under past democratisation efforts as utopian and absurdist:

> If we set ourselves the objective of creating some sort of central Asian Valhalla over there we will lose because nobody in the world has that kind of time, patience or money to be honest... It seems to me we ought to keep our objectives realistic and limited in Afghanistan... Otherwise we will set ourselves up for failure. (Spillius 2009: 14)

Gates noted the lack of political will as a fundamental obstacle blocking the goal of liberal stabilisation. In mocking the Bush Administration's (and, by proxy, the Blair government's) desire to create a modern, liberal democratic 'central Asian Valhalla', Gates set in motion a popular rhetorical device that would be imitated by members of successive British Governments to attempt to create distance between their own revisionist goals and those declared at the height of Blair's tenure. Ministers would frequently use their imaginations to juxtapose some Western locale with that of the impoverished Afghans in order to downplay the human security and democratisation elements of the stabilisation process; some of the more inventive ways of communicating official disinterest in democratisation include claiming Britain is not in Afghanistan to 'create Hampshire in Helmand', 'Switzerland in Afghanistan', and 'the new Jerusalem in the Hindu Kush' (Simpson 2012: 123; Watt and Wintour 2010: 13). This is worth noting because it demonstrates two points. Firstly, it shows the extent to which British government took its lead from the Americans not just in terms of matters of policy and strategy, but also in terms of how those policies or strategies are conveyed, even down to creating geographically- and culturally-tailored variations of off-the-cuff idiosyncrasies.

Secondly, it reveals once more the impact of the Iraq Surge on the mentality of policymakers on both sides of the Atlantic, for it was during this period that the revisionist technique of *reductio ad absurdum* originated in an opinion piece by public relations consultant Tim Hames, who argued Iraq would never become 'Sweden with beards' (Hames 2007: 19). It is, of course, undoubtedly understandable that this approach would become common currency amongst political leaders, given the stuttering impact of stabilisation efforts in Afghanistan by early 2009 and the all-consuming imperative of achieving a semblance of security in the country. In this respect, the *reductio ad absurdum* narrative technique can be seen as an eminently sensible, if somewhat crass corrective for the erstwhile pervasiveness of universalistic claims to western-centric normativity. Narrative changes did not change the operational realities of Afghanistan, however; the entire stabilisation process remained predicated on a developmental and governance initiative founded on the Afghan Compact of 2006, which called for nothing less than a complete transformation of Afghan social, political and economic life within just four years. Thus, whilst ministers were labouring within a framework expressly committed to creating such a grandiose and utopian eventuality, they were simultaneously undercutting it by deliberately and concertedly downplaying its feasibility.

Within Whitehall, Obama's new strategic approach would have been welcomed by most, not least within the Ministry of Defence which, under Hutton's leadership, had set in motion a more circumscribed account of the Afghan mission the previous November. Whilst remaining true to the overarching democratisation element within the stabilisation narrative by referring to Britain's role in 'securing the Afghan democracy', Hutton also distinguished himself from Miliband's by then long established mantra that 'there will not be a military solution' to the conflict by appearing to confirm interviewer Jon Sopel's suggestion that the international community could 'militarily beat the Taliban' (Hutton 2009, online). This chimed with the operational approach on the ground in Afghanistan at the time, which was 'focused on defeating the insurgency' in a 'fairly conventional military campaign' (Chaudhuri and Farrell 2011: 272–3). Even Miliband, the intellectual descendant of Blair's 'universal values' mantra, went to some effort to reduce his insistence on human security and democratisation by arguing in greater force for the centrality of national security and against the cavalier tendencies of Blairite liberal interventionism (Miliband 2008e). Indeed, according to journalist Richard Norton-Taylor (2009: 7), by mid-January it was the popular opinion (privately uttered) within Whitehall that ministers and officials would be better placed in attempts 'to get their message across [of the importance of the war in Afghanistan to the British public] if the emphasis is placed on Britain's interest rather than on improving the lives of Afghans'. With this in mind, it is telling that by the end of January it was reported that the Cabinet Office was seeking to hire several 'strategic communication' practitioners to, as one journalist interpreted it, assist in the 'massaging' of bad news (Walker 2009: 8). It is almost certain that this shift in both emphasis and method of delivery of the stabilisation narrative was in place across the whole of Whitehall by the end of February, and that it was the implementation of Defence-led SC practices within Whitehall that cemented the national security angle to the stabilisation policy narrative.

In addition to this new approach essentially representing a victory for the Army and MOD's SC work (both in terms of consolidating messages across Whitehall and of repositioning communications in a manner that reflected the sacrifices of the military), such an interpretation represents a succinct articulation of the Government's desire to ignore the politically unstatable realities of the transnational dilemma in favour of a realist-centric approach to communication. This new approach would be compromised by *The Telegraph* the following day, however, in an article that argued that a prevailing sentiment within Whitehall was that 'we should cut our losses and get out' of Afghanistan, and that reasons of alliance dependency were dictating both the narrative revisionism taking place and the lack of dissenting voices within the political establishment:

> a depressing uniformity of outlook prevails among politicians in the two major parties, namely that a critical view of what is happening in Afghanistan might undermine the western alliance. Our young soldiers are being killed just to show willing in Washington, doubly so now that a popular Obama has replaced Bush. (Burleigh 2009: 22)

The transnational dilemma was becoming as politically toxic to communication efforts as Blair's universalistic rhetoric had been several years prior. Again – and this is a point in need of reiteration – the problem of explaining the relevance of collective security operations honestly and accurately is that one cannot do so simply and quickly. To do so would require the conceptualisation of a largely indirect threat in Afghanistan in a manner that appeared immanent and direct. Likewise, to seek to explain, simply and quickly, the dynamics, expectations, and responsibilities of membership to a collective security framework will necessarily incur a good deal of omission. The SC efforts made by Government officials to reconstitute in the British public's consciousness the purpose of Afghanistan as foundationally a national security issue – where Afghanistan was a means to an end of British security – was useful and necessary to overcome both the international issues of carrying a disproportionate amount of the burden of stabilisation and the domestic institutional fallout associated with issues of obligation and cooperation. However, by gutting the normative aspect of the mission – that which constituted the institutional make-up of NATO and thereby made the stabilisation strategy possible – the stabilisation narrative was also substantively compromised by narrative reconfiguration. This in turn led to doubts amongst some about the supposed necessity of Afghanistan to British security in the face of equally prescient risks elsewhere. Afghanistan did represent a threat to Britain's national security, but no more so than Somalia or Yemen and certainly less so than neighbouring Pakistan. Rather, the importance of Afghanistan to the national interest and security of the United Kingdom remained the same as it ever did, and certainly the same as it did since NATO acquired ownership of ISAF in 2003: to act as a proving ground to bind the members of NATO together in order to ensure the survival of the pan-Atlantic collective security mechanism, *through which, and only through which*, Britain's security needs can be delivered. Such a reading provides an honest appraisal of why Afghanistan is important; this is not, however, the same as arguing a proven link between British operations in Afghanistan and a verifiable diminishing of Afghan-based terrorism against the United Kingdom.

This, then, was the purpose of British SC practice: to square the circle between the mundane realities of collective security responsibilities, on the one hand, with the public desire to be told that the losses incurred by the state in blood and treasure in Afghanistan were of some immediate and direct utility to national security. To put it another way, SC sought to avoid the potential for critiques that could conceive of British sacrifices in Afghanistan as a regrettable but necessary kind of collateral damage in pursuit of maintaining Britain's place within the international community. Remarkably, however, it appears that the feud between Downing Street and the Armed Forces would work to mitigate the effectiveness of SC efforts from practically the outset. In what one might see as the second round of Brown versus Dannatt, the Chief of the General Staff renewed hostilities by allegedly making an offer, in conjunction with Hutton and allegedly without the knowledge or approval of Brown, of 2,000 additional troops to the Americans on the eve of Obama's 'Af-Pak' policy announcement (Evans and Coates 2009: 1). Indeed, just a few days later, Dannatt would take part in an interview with *The Times*, in which he undid SC efforts by citing the unstatable transnational dilemma in claiming that the *quid pro quo* of alliance negotiations was the reason Britain entered Helmand, and posited that then-Chancellor Brown had been reticent to provide the necessary financial resources for the mission (Evans 2009: 30; Usborne and Merrick 2009: 14). In one fell swoop, rightly or wrongly, Dannatt had brought Brown's reputation into disrepute and undermined the national security argument in order to defend the Army's interests.

Despite British efforts to downplay the disadvantages of collective security membership, events in the first months of 2009 indicated that alliance cooperation was indeed a central issue for all parties concerned. Gates expressed dismay as to the poor likelihood of securing greater commitments in troops from America's continental allies, and rumours circulated within the British press that the Obama Administration was unimpressed by the progress made by British forces in Helmand, to the point where one journalist argued that American forces in Helmand 'refused to take orders' from the British there (*The Guardian* 2009b; Starkey 2009: 24). Ostensibly due in parts to the pressure exerted from the Armed Forces to either materially support the mission as it stood or scale down their ambitions, demands from the Treasury to not stretch public funds beyond breaking point, narrative revisionism from both the Americans and from within his own Cabinet (courtesy of Hutton), and the humbling effect of having British leadership in Helmand compromised by the assertiveness of the American contingent, Brown took the step in late April to reframe the mission in distinctively counter-terrorist phraseology, typified by his rhetorical labelling of the Af-Pak theatre as the 'crucible of terror'. He informed the Commons that, rather than democratising Afghanistan as a worthy goal in itself, Britain's 'strategy is to ensure that the country is strong enough as a democracy to withstand and overcome the terrorist threat' (Hansard 2009b). While still nominally asserting the validity of the liberal peace thesis, the stabilisation mission in Afghanistan was presented in its starkest terms yet as almost purely about British security needs and, as a corollary, essentially marked the end of the democratisation pathway of the stabilisation narrative (Wintour 2009: 8). Despite his efforts to reframe the mission in minimalist, security-centric terms, Brown still incurred heavy criticism from the usual sources, with *The Telegraph* and the Conservative opposition arguing that, in failing to agree to the 2,000 troop increase advanced by Hutton and the Armed Forces, he had damaged Britain's reputation with the Americans and had 'betrayed' the Armed Forces (Coughlin 2009: 30). The double bind of transnationalised defence policy remained: balancing the obligations made to collective security partners with those made to the military seemed to be an insurmountable task. At the end of the first week of June, following a failed Blairite putsch against his leadership, Brown had led the biggest Cabinet reshuffle of his tenure and accepted Hutton's resignation, presented as resulting from 'family problems' (Porter and Adams 2009, online).

By the end of the summer, emphasis on national security informed a stabilisation policy narrative that had become a hollow facsimile of its former self and, consequently, was practically unrecognisable from the aims and ambitions of its original position in 2006. The civilian goal of entering Helmand – to pursue a reconstruction and development agenda as a counterbalance to the counter-terrorism focus of the American forces – had been reduced to almost an afterthought in narrative terms. With the decision to reduce the area of operations of British forces from the entirety of Helmand to a smaller zone within the centre of the province in July, so too had the ambitions of the generals – to redeem Britain's military reputation post-Iraq by pacifying Helmand and defeating the Taliban – been rendered obsolete (Griffin 2011: 318). Coughlin (2009b) would lament this decision as resulting in Britain finding itself 'once more in the humiliating position where we are having to rely on the Americans to do the job'. Indeed, to the extent that the Americans had saved the British in the latter, so too had they negated the desire for British autonomy in the former, for the British operational plan was rendered subservient to the strategic direction of the Obama administration. It is clear that by this point the process of narrative articulation started by Hutton in late 2008, and cemented by Obama in early 2009, was building a self-perpetuating momentum within the British state: narratives were reframed, resources were reallocated, strategies were reordered, and ambitions were contracted.

This represented a complete reversal of the trend established from 2002 to 2008, where the stabilisation approach and corresponding narrative had expanded almost exponentially to cover the tertiary requirements (reconstruction and development) for a secondary objective (stabilisation) to allow for a primary goal (counter-terrorism) to be achieved. It is perplexing then that, in making the case for the counter-terrorism approach and for Britain to continue adhering to its commitments in Afghanistan, British officials would tend to undercut their own argument. An ideal example comes again from Dannatt, who spoke at a Royal United Services Institute lecture in June 2009 of the need to 'do whatever is necessary' in Afghanistan because 'success' there was 'not discretionary' (Norton-Taylor 2009b: 26). Of course, the point Dannatt was trying to make was the centrality of stability in Afghanistan to Britain's national security, which should be considered non-negotiable and beyond compromise. In reality, however, success in Afghanistan by this point was wholly discretionary, since the vast array of targets and objectives that were originally conceived as necessary prerequisites for success were no longer given much importance, and instead were replaced – as a matter of discretion based on resources available and feasibility – with a counter-terrorism approach that favoured realistically achievable goals. In fact, even the term 'success' was discretionary: whereas wars (including this one) are usually understood in traditional terms of victory and defeat, the lexicon for protracted counter-insurgencies is now one of 'success or failure', indicating, once more, that success itself is a term discretionarily revised down from the original narrative of victory (Snyder 2011: 25).

Putting issues of semantics aside, the disunity in civil-military relations had reached new heights by the end of the summer, and the arguments therein – focusing on equipment and helicopter levels – gave an impression that there was not universal acceptance of the crucial nature of the mission in Afghanistan or, if there was, that the means by which Britain could assure its own security was indeed discretionary. This was almost certainly Dannatt's point, however: that because the mission went to the heart of Britain's security interests, Government should therefore not spare any expense in providing the military with everything they say they require. Transnational policy dilemmas emerged once more: Brown was again pressed by the Americans to increase troop contributions whilst also refusing to meet the demands of the service chiefs, the chiefs once more circumvented his authority by going to the press, stating that additional helicopters and equipment were necessary to save soldier's lives, thereby delivering Brown with an ultimatum of either acceding to their demands or risking the public's wrath by being seen as indifferent to the safety of British troops (Sengupta and Morris 2009: 10; Hinsliff 2009: 18). The obligation-cooperation cycle was out of control and the tensions that it produced were undermining Government credibility. The collective security obligations imposed upon the United Kingdom – which was by this point 'running hot' in terms of both its military capabilities and its budgetary expenditure, was having the institutional effect of breaking the fabric of civil-military relations, and therein incurring a knock-on effect on the ability of the state to communicate effectively and in unison on the purpose of and reasoning informing continued participation in Afghanistan. In an interview with a former official, the equipment 'scandal' was presented to the author as the *coup de grâce* to communication between elements within the military community and the Brown government:

> Richard Dannatt, then chief of the military, actually argued that this would be a good opportunity for good soldiering. He was a firm advocate of the strategy and when the going got tough he was nowhere to be seen actually justifying the rationale for it. (Personal Interview 2013)

Dannatt's fervent advocacy of the Army's interests was having severely deleterious effects on British messaging. For Dannatt's part, his decision to go 'off-message' was driven by what he perceived as his responsibility to defend the interests of the Army against a Government he viewed as riding roughshod over the Military Covenant. Indeed, this was a long-standing position for Dannatt, stretching back to 2006–2007; in his autobiography, he claims that 'national leaders seemed unprepared to speak up for' Afghanistan in the face of increasingly negative media reporting, and that it was from that point that he decided to pursue his own communications agenda (2011: 354–5). He claims that it was his outspokenness on Afghanistan – in direct contravention of Government's agreed lines – that precipitated a highly personal attack on his character by Labour ministers in the months leading up to his departure from his position as Chief of the General Staff in 2009 (2011: 27–8). Despite the Labour Government's misgivings about what they perceived as Dannatt's unprofessional conduct, elements of the national press feted him as a courageous advocate for an underappreciated Armed Forces (Grice 2009: 17). Whatever the virtues of Dannatt's actions, the rift between political and military leaders over the summer of 2009 – the deadliest for British personnel since the conflict began, with 65 deaths between May and September – demonstrated the difficulties of maintaining a coherent narrative (and through that, sustained public support) for Afghanistan in the absence of unified leadership and the presence of severe institutional in-fighting.

The one area of convergence highlighted by all state representatives by the autumn was the centrality of the mission to British national security. Indeed, the damage done to the stabilisation narrative likely made national security the only common ground left. As such, all non-military and non-security aspects of stabilisation in Afghanistan were downplayed or openly disparaged as excessive, unrealistic, or secondary. This focus on British interests and security developed to the extent that ministers openly disregarded the objective of improving Afghanistan for the Afghans' sake as a means of emphasising the core goal of British security. Then-Armed Forces Minister Bill Rammell spoke of continuing the mission 'not because we want to make the world a better place in that part of the world', while Miliband, for whom universal values and democratisation were once so central, made the point that 'the government has committed our troops to Afghanistan not for Afghan democracy but for our security' (Miliband 2009b, online). There were aberrations to this trend, of course: for example, the build-up to the Afghan presidential elections in August provided some within the Labour Government an opportunity to renew the goal of democratisation (notably Denis MacShane, who claimed those who advocate withdrawal from the conflict were 'defeatist[s] who care nothing for democracy, human rights or the need to send a "No Pasaran" signal to those who hate democracy' (MacShane 2009)). Such aberrations in messaging would prove short-lived, however, as the election was widely perceived as being compromised by fraudulent activities, much of it levelled at Karzai supporters. By the end of 2009, Shadow Defence Secretary Liam Fox issued a call for an end to democratisation efforts (rather superfluously, given it had all but disappeared by this point anyway) and advocated a further narrowing of goals, ostensibly presented not as a failure of liberal universalism but as a kind of anti-colonialist celebration of differences:

> we've got to stop judging Afghanistan by Western standards. We're not trying to apply a Jeffersonian democracy to a 13th century state. If we try that, we'll be unsuccessful. I think we've got to stop trying to make judgements from outside about what's good for the Afghan people. What we want to see is a stable enough Afghanistan, so it's not a risk to our national security. (Fox 2009, online)

Indeed, whilst the narrative trend from within the Brown ministry had been to advocate much the same throughout its duration, their language was understandably more measured given the diplomatic responsibilities of power. Interestingly, in the face of criticism of both its support for Afghan democracy and of allegedly turning its back on Afghan democracy, Labour ministers spent the remainder of Brown's period in power downplaying the expectations for lasting democratic governance or human rights protections in the country. Rammell, for example, would argue for a 'reality check' on overly optimistic expectations for the 'fifth poorest country in the world that doesn't have a tradition and a history of democracy', while then-Defence Secretary Bob Ainsworth, when confronted on television by a fellow panellist for failing Afghan women, argued that 'Afghanistan will always be a very different country to our own and will have different values, and we are not there to create some eastern version of Great Britain' (Rammell 2009b, online; Ainsworth 2010, online). Given this trend under the leadership of Gordon Brown, the Conservative-led Coalition Government could continue the pared-down stabilisation narrative in a near-seamless transition. It is unsurprising that this should be the case precisely because the Labour process of narrative deflation – highlighting national interest and security above all other aspects – begun in earnest by Brown in 2007, and with renewed vigour by Hutton in 2008, were actually Conservative policy positions in the first place, dating back to David Cameron in 2006 and Fox in mid-2008. In this sense, then, the Coalition was well placed to advance further the narrowing of the aims of the mission once in power. Indeed, they wasted little time in doing so, with Cameron announcing, just over a month after becoming Prime Minister, his intention to begin a timetable for withdrawal, or 'drawdown', of five years at the G8 summit in Toronto in June (Wintour 2010: 1). According to one former official, this announcement came as a total surprise to all and was never discussed in Cabinet or the National Security Council, nor was the Ministry of Defence given advanced warning (Personal Interview 2013). It would appear that while the decision was taken, as one former official said, 'pretty much unilaterally' by Cameron in the United Kingdom, he was also acting in concert with Obama, with whom Cameron held a pre-Toronto meeting and, like Cameron, held the view that, in the wake of ISAF commander General Stanley McChrystal's sacking just days prior to the meeting, the operational approach of 'population-centric counterinsurgency' was not producing results quickly enough. Indeed, much like the United Kingdom, the Obama Administration was riven with division over the best approach to drawing down forces in Afghanistan and whether counterinsurgency was the most suitable option for American strategy in the country (Sanger 2012: 34–8; Hastings 2013 134–5).

Cameron's decision to place a timetable on drawdown irrespective of events created yet another narrative discrepancy within Whitehall, placing him at odds with military chiefs, who opposed the idea, and Fox at the Ministry of Defence, who remained adamant that a timetable would be counter-productive and allow the Taliban to simply wait ISAF out, as well as representing a 'shot in the arm to jihadists everywhere' (Watt 2010: 13). It would be nearly a fortnight before Fox would eventually step into line with the timetable plan, at which point he argued that '[t]here has always been a timetable', quite at odds with his previous statements (Kirkup 2010, online). Once it became clear within the international community that Britain, for so long among the most committed of major European states to the stabilisation process and among the most willing to carry the burdens of an unbalanced alliance and, of course, the convening nation for that process begun in London in 2006, was making haste to depart Afghanistan, the long-term stabilisation programme Blair and Brown had advocated quickly disappeared. The Dutch and Italians announced their withdrawals, and think tanks such as the International Institute for Strategic Studies and various academics increasingly advocated a pared-down counter-terrorist approach, including the use of unmanned aerial 'drones' undertaking 'targeted assassinations' in the Af-Pak tribal areas (Cassidy 2010: 42; Blackwill 2011: 46–8; McCrisken 2011: 797; Hudson, Owens and Flannes 2011: 122–32; Belcher 2012: 260).

The timetable process was formally agreed by NATO member states at the November Lisbon Summit; by the end of 2010 the 'whole thing was pretty well done and dusted' (Personal Interview 2013). Running alongside this rapid period of change was a similarly hasty attempt to further revise the stabilisation narrative for public consumption. The Coalition's staple narrative for Afghanistan had now taken on the form of a more forceful denial of an ongoing role for democratisation in the stabilisation process, with ministers all producing some variation of the comments that 'we are not in Afghanistan to build a perfect democracy' or 'a carbon copy of western democracy or convert people to western ways' (Cameron 2010; Hansard 2010d; Fox 2010b, 2010c; Harvey 2010). Peculiarly, for some in the Labour Opposition who followed suit, this involved denying democratisation had ever been a core element of the stabilisation campaign; Ainsworth declared in a September 2010 interview on the future of British operations in Afghanistan that 'no one has ever believed that you can create a Western-style democracy in Afghanistan' (Ainsworth 2010b, online). Labour's narrative approach of distinguishing between western and non-western cultural traditions is problematic, however, since their Kantian worldview – the philosophical bedrock of democratisation in Afghanistan – rejected the very idea that democracy should be understood as a culturally-specific phenomenon. In the face of mission objectives that could not be met, to which the leadership of all parties had at the time pledged their support, the politicians of all parties chose to close ranks by opting to rewrite the history of the stabilisation campaign to one that fit with the only narrative left by which success could be found.

Phase V: 2011–2014 – Conclusion: Drawdown and National Interest

The trajectory of the stabilisation narrative from internationalist liberal rhetoric to narrowly-defined realism was all but complete by the beginning of the fifth and final phase of the Afghanistan campaign, and the killing of Osama bin Laden in May 2011 provided the Obama and Cameron governments a useful foil for measuring progress in the country and justifying the narrowing of objectives and ambitions between 2010 and mid-2011 (Dodge 2011: 69). In June, Obama announced the beginning of the end of the 'surge' and the withdrawal of 33,000 troops by the end of 2012 (Obama 2011, online). The day after Obama's announcement, Cameron indicated a similar intention to withdraw British troops in an incremental fashion; by the end of 2011, Cameron would put forward proposals to advance the drawdown to nearly half the British contingent by the end of 2013 (Kirkup and Spillius 2011: 1, 2; Hopkins 2011: 1). Given such clear political signals, the United States' and Britain's allies duly followed suit, and the period 2011–2014 witnessed the disorderly acceleration of drawdown timetables by several NATO member states, notably the French in January 2012. Dorman (2012: 303) noted that the months leading up to the May 2012 NATO summit in Chicago were marked by a failure on the part of alliance members to agree on an alliance-wide exit strategy. With the British and the Americans no longer pressing the case for a 'conditions-based' withdrawal and, on the contrary, leading the way for withdrawal by introducing an effectively non-negotiable timetable process grounded in national security interests, it appears that the institutional and normative ties that bound lesser European states to the Afghan mission quickly unravelled. The events of Lisbon and Obama's post-bin Laden speech had effectively opened the floodgates for a disorderly 'dash for the exits' (Dorman 2012). It was in this context that the

Chicago Summit was convened to mitigate the worst effects of the dissolution of political will for continuing on in Afghanistan. The result of the Summit was a further speeding up of the drawdown process, with the end-date for combat operations pushed forward from the end of 2014 to mid-2013 and a focus on 'transitioning' with and training Afghan security forces between 2012 and 2013.

The accelerated nature of the coalition drawdown from Afghanistan was undoubtedly the result of a loss of political desire to carry out the stabilisation process as envisaged by Brown three years earlier. Brown's multi-dimensional, long-term vision for the rehabilitation of Afghanistan's civil society and economy was deemed unrealistic by his political opponents, who managed to win skirmishes with the Labour leader over the nature of the mission through various means, not least of which by pointing out the dangerous degree to which the original objectives had spiralled outwards as a result of the 'comprehensive approach'. By reducing the scope of the mission and reframing its purpose as primarily – and eventually solely – about Britain's national security and national interest, the Brown and Cameron Governments managed to correct the course of the strategy and, in doing so produce a policy narrative amenable to a gradual retrenchment of core objectives and capable of deflecting criticism resulting from the inevitable shortcomings of Afghan democratisation efforts. The view within Whitehall that gaining and maintaining public support required such a transition allowed the narrative and strategy to synchronise to a degree unseen since the initial weeks of the Helmand mission, and in doing so temporarily improved the prospects for sustaining the mission until the core task of stabilising the central government and its security apparatuses had been nominally achieved. However practical and politically necessary this course of action may have been, it had the result of an unfortunate no-win scenario of choosing between two unattractive pathways: to use a medical analogy, doing nothing to stem the infection of a narrative out of sync with events and risk total collapse, or taking action to stop the rot through partial amputation of the narrative and invariably risking losing the patient altogether. Britain and its allies chose the latter, and in doing so undermined the preceding efforts and thereby called into question the point of the stabilisation mission altogether. A narrative focus on national security as a means of avoiding the political unpleasantness of the transnational dilemma *in order to sustain the alliance and to sustain the coherence of the British state* ironically contributed to the dissipation of alliance unity by inculcating the conditions whereby the primacy of collective security to Britain's interests in Afghanistan could be essentially ignored in favour of individual domestic political concerns.

In broad terms, the Coalition Government did achieve the core tenet of SC doctrine as outlined in *JDN 1/12* by realising the synchronisation of word and deed: they successfully moved the stabilisation policy narrative from its disconnected heights of Helmand in 2006 and early 2007, when what was being said about Afghanistan bore little to no relation with what was actually happening there, to one where communication about the conflict was more or less equivalent to the ground truth. Solving this dilemma is not to be underestimated, but it invariably came at a cost by creating a whole new dilemma which was partly ethical and partly practical. The paradox of SC in Afghanistan is that the reduction of strategic goals, initiated as it was by Hutton's 2008 speech, came at the expense of the wider strategic coherence of the intervention as a whole. In other words, the need to explain the conflict clearly and succinctly led to a simplification of the institutional and normative basis of the strategic paradigm for the conflict itself; but because the strategy was progressively simplified and cleaved away from that point, the narrative no longer made sense and exposed new contradictions as a result, effectively bringing the utility of the stabilisation policy narrative full circle to return it to its original point of incompatibility with strategic necessity. By late 2013, Cameron would offer to assembled media his opinion that efforts in Afghanistan had been successful, but only according to the metrics his Government had set out:

> Here we are in Helmand Province. There was no Afghan National Army. There were no Afghan National Police. There were no women going to school, girls going to school. We see a situation totally transformed... I think it is good enough. It is not perfect, it is still a very poor country, it is still a democracy that has a lot of development yet to do, but our aim has always been, can we make sure that this country is no longer a haven for terror and terrorists. I think we have done that, and in the process we have trained up an Afghan National Army and an Afghan National Police force that are capable of delivering the basic security this country needs... I think our troops can leave with their heads held high over a job very well done. (Cameron 2013, online)

In giving little credit to the narrative focus on the comprehensive approach as essential to Afghan stability and global security in favour of a minimalist security-centric approach, Cameron and the Coalition left themselves open to accusations of abandoning much of the substance of the mission, such as democratisation, human rights and the advancement of women, the consolidation of a liberal Afghan constitution, and various other aspects of the Afghan National Development Strategy as enshrined at the 2006 London Conference. In addition, the Coalition's narrative revisionism left it vulnerable to recriminations relating to the argument that Afghanistan was crucial to maintaining the collective security framework: the accelerated drawdown was the culmination of years of faltering attempts to unify and cohere NATO member states and their armed forces into an efficient and well-structured security umbrella capable of meeting twenty-first century challenges of the type represented by Afghanistan (Dorman 2012). What is more, the long-standing rationale that guided the slow but steady expansion of mission objectives and sub-objectives in Afghanistan – that a continued presence was necessary lest the jihadist opposition be convinced the West had capitulated – appears to have been completely undermined by the rapidity of NATO's drawdown process. This may explain Fox's (indeed, the vast majority of all British policymakers and military leaders apart from the Prime Minister) apparent reticence to agree to the timetable process advanced by Cameron, in case such extremists receive 'a shot in the arm' (Hansard 2010d). Finally, and perhaps of greatest significance for the future of western power, is the effect drawdown may have on the institutionalisation and normative legitimacy of liberal interventionism. That the drawdown process took place in parallel with the so-called Arab Spring is worth noting, for while the stated objectives for Afghanistan by this point contained little if any reference to democracy, the popular uprisings across the Arab world appeared to briefly give succour to the Kantian streak in some politicians. For example, Liam Fox's rhetoric on the removal from power of Libya's Muammar Gaddafi was reminiscent of Blair and Miliband when he opined that

> the Arab Spring and what NATO has been helping to achieve in Libya is so important. It shows that violent extremism is not the only route to change. It demonstrates that representative government and freedom are not simply 'western' values but represent a universal aspiration. (Fox 2011b, online)

As was the case with Afghanistan, however, events in the Arab Spring would rapidly undermine this return to the democratising streak in British policy narratives however. As Katerina Dalacoura (2012: 71–6) recently noted, the revolutionary fervour across the Middle East may have produced calls for democracy, but the notion that democratisation would result in anything like the western liberal variety would be 'wrong and simplistic'; indeed, time and again this has since proven to be the case, with popular uprisings in Egypt and elsewhere producing Islamist governments and, as Dalacoura asserted, more often came about as a result of widespread animosity to perceived neo-liberal hegemony than in support of Western norms. Additionally, the West was remarkably selective in its support for revolutionary movements during this time: while Egyptian protesters were (belatedly) offered solidarity and Libyan rebels were provided with NATO military support during the civil war against Muammar Gaddafi's forces, movements in Bahrain and Yemen were regarded with ambivalence or hostility. The ongoing civil war in Syria between President Bashar al-Assad and a complex array of forces has produced even more controversy, for while Britain and the United States have offered considerable support for elements of the anti-Assad insurgency, they failed to secure popular or parliamentary mandates for military intervention and, in attempting to do so, appear to be at odds with the stabilisation policy in Afghanistan – that is, seeking to shore up one regime that was tasked with fighting Islamist insurgents whilst simultaneously attempting to destabilise another that was doing much the same thing.

As it were, the general political discomfort on display during Cameron's fateful decision in August 2013 to seek parliamentary consent for increasing Britain's role in the Syria affair signalled a distinct loss of appetite for military adventurism in general, and also indicated that Cameron's narrative revisionism in Afghanistan may have created blowback on Britain's wider foreign policy aims: by demarcating the objectives in Afghanistan to those solely concerned with

British national security interests, he may have undermined the appeal to intervene in Afghanistan on purely normative grounds of preventing the use of chemical weapons on Syrian civilians (Personal Interview 2013). Loss of political appetite to intervene on the scale required for the comprehensive approach was also evident in the Libya campaign of 2011 and the campaign against the 'Islamic State' in Iraq in 2014, in which coalition forces restricted themselves to maintaining no-fly zones and air strikes and explicitly ruled out the use of large-scale ground forces. This lack of stabilising presence can be pointed to as a possible causal factor in the resultant destabilisation of neighbouring North African states, notably Mali, as insurgent forces were freed from the restrictions imposed by the Gaddafi regime.

Conclusion

By reconfiguring the stabilisation narrative in such a wholesale fashion, the Brown and Cameron Governments risked the obsolescence of the entire rationale for Britain's presence in Afghanistan since the NATO takeover and expansion of ISAF in 2003. This is, of course, a distinct irony given that reconfiguring the narrative was necessary for shoring up domestic and international public and political support for the mission, in order to prolong it to the point where a reasonable and responsible withdrawal was possible (Jones and Smith 2010: 115). Ultimately, the historical record will (and increasingly already is) coming to a critical conclusion regarding the stabilisation mission as a regrettable waste of time, resources and lives. If, as the stabilisation narrative from 2008 to 2014 suggested, the goal of the United Kingdom and its coalition partners had been as minimal as Cameron's quotation above argues – that of a stable (not necessarily democratic, at least as far as it is understood by Western peoples) Afghan state and an Afghanistan that does not provide a 'safe haven' for terrorist organisations – the question naturally arises of whether bothering with much of the work of stabilisation was necessary in the first place. Since the objectives of the mission were pared down to equate, in historical terms, with those that preceded the inception of the stabilisation narrative in mid-2002, one might critically conclude that the majority, if not the entirety, of the stabilisation narrative developed from 2002 onwards may be called into question. To indulge in one hypothetical, it is just as possible that the minimalist objectives emphasised by the Coalition could have been negotiated in 2002 with remnants of Taliban as it is to posit that to do so would invite insecurity for the United Kingdom and its allies. Equally, it is far from evident that the counter-factual exercise of predicting the negative effects of leaving Afghanistan in a morass of post-conflict instability in 2002 would have been more deleterious to British national security and interests than the negative effects of staying in Afghanistan until 2014.

In jettisoning the non-security elements of stabilisation to the margins of narrative, these questions are actually given greater force; indeed, it is uncertain whether the NATO alliance can organise itself for stabilisation operations in the future without a core democratisation ethic (Birkle, O'Hanlon and Sherjan 2011: 11). The one common thread upon which stabilisation efforts hung – albeit so precariously throughout – was the grand ideological vision of improving Afghanistan as not just an end for global or national security but as a morally worthwhile exercise with the end of improving the prospects of the Afghan people. Without this core ethic of liberal interventionism, the realist, nationally-oriented manifestation of the latter stages of the stabilisation policy narrative has made itself vulnerable to the same self-defeating relationship between communication and strategy that became so apparent under Blair. However, it would appear that it was simultaneously necessary reconfigure the stabilisation narrative in such a way to heal the rifts that the stabilisation policy's entropic expansion had caused within the political and military establishment of the United Kingdom. Although distinctly limited in in its actual policy options to this end, as a consequence of the prevailing necessities of maintaining alliance cohesion and its relationship with the United States, and of preventing the total implosion of civil-military relations in the eyes of the public, the paring down of the stabilisation narrative provided the Brown and Cameron Governments with the discursive pretext for the eventual transition to a more limited version of stabilisation that was, in time, compatible with the over-riding objective of maintaining its political and military obligations to the United States. The same dynamics are evident in the next chapter pertaining to the other normatively-based aspect of UK operations in Afghanistan: counter-narcotics.

Chapter 5

The Counter-narcotics Narrative

In many ways, an account of Britain's counter-narcotics narrative appears to be simply an adjunct to the wider stabilisation narrative outlined in the previous chapter. Indeed, counter-narcotics arose out of the same philosophical and institutional milieu as the stabilisation agenda. The assumptions informing the development of counter-narcotics policy in Afghanistan fit the criteria of liberal peace theory insofar as the drugs trade represented a challenge to the development of liberal democratic governance. Those involved in the manufacture and trafficking of opium were considered to be malignant social actors comprising an alternative locus of power to the Kabul Government's officials, and so combatting their activities could be reasonably seen as a means of promoting a more accountable system of administration and the growth of licit economic activities. Pursuit of counter-narcotics in Afghanistan also fulfilled other aims implicit within stabilisation. It coincided with New Labour's ethical foreign policy by focusing on the evils of heroin use on the streets of Britain. Moreover, counter-narcotics held ostensibly strategic aims for global security, as the opium economy was seen as a considerable source of revenue for al-Qaeda and the Taliban. Perhaps most importantly, Britain's role in counter-narcotics operations in Afghanistan fulfilled its own strategic objective of maintaining the alliance system by contributing not only to Britain's 'soft power' development agenda but also, crucially, by 'showing willing' to the alliance by performing a task assumed to be beneficial to the US-led War on Terror but which the Americans were themselves reticent to participate in.

Despite the commonalities between counter-narcotics and stabilisation (and to some extent counter-terrorism), the counter-narcotics policy narrative is distinct and independent from those of stabilisation and counter-terrorism in one important respect: unlike with the counter-terrorism and stabilisation policy narratives, counter-narcotics policy was the only aspect of Britain's efforts in Afghanistan that was at any point within the near total control of its government. Curiously, as of the end of 2014, it is also the one area of policy that effectively disappeared from official statements on the purpose of British participation in the conflict. What the evolution of the counter-narcotics policy narrative demonstrates most clearly is that the abiding realities of Britain's institutionalised acquiescence to the demands of collective security prevented it from maintaining control over its own policies in Afghanistan, thereby rendering the articulation of consistent messaging on said policies a near impossibility. This is made most obvious in the case of counter-narcotics because it is one where policy was, at some point, nominally within the control of Government and where its rhetoric and accompanying strategies was devised with large degrees of autonomy from (but still occurring within) a multinational framework. Britain's ownership of the counter-narcotics policy narrative can be seen as a result of its lead nation status on the issue within the international community, which is itself a reflection of a distinct lack of interest in counter-narcotics on the part of the United States for much of the Afghan conflict. The more American interest in counter-narcotics grew (and ultimately dissipated), the more Britain's policy narrative had to alter to remain in tune with that of its senior partner. In other words, the relative independence of British policy formulation on counter-narcotics eventually had to be reconciled with external factors related to collective security dynamics – specifically pressure from the United States on the United Kingdom to modify and eventually drop its counter-narcotics policy to reflect Washington's point of view – thereby ultimately causing the policy narrative to lose its coherence over time.

Phase I: 2001–2003 – The Domestic Policy Origins of the Counter-narcotics Narrative

In order to understand how the United Kingdom came to be responsible for the seemingly Sisyphean task of dismantling and destroying the Afghan opium economy, we must first say a few words about the Blair Government's views on drugs in Britain. New Labour's drugs policy was at the heart of its much publicised 'tough on crime, tough on the causes of crime' rhetoric established during its time in Opposition during the mid-1990s. Central to this approach was the relocating of drug offences within a wider context of anti-social behaviour and criminality symptomatic of the 'wreckage of our broken society', as Blair put it (Blair 1995, online). This was something of a departure from earlier approaches that sought to distance drug addiction from interdiction efforts and instead place it firmly within the confines of a social health issue (Stimson 2000, online; Hunt and Stevens 2004). New Labour's accession to power in 1997 was followed by a White Paper entitled 'Tackling Drugs: Where We Are Now' in April 1998, the content of which focused on the social, psychological, economic, criminal, and health-related 'evils' of drug use and promoted a vision of a 'healthy and confident society increasingly free from the harm caused by the misuse of drugs' (HM Government 1998: 9). Indicative of the 'causes of crime' mantra espoused a few years earlier, the Paper outlined the need for a comprehensive approach to drug policy that cut across domestic and foreign policy boundaries. Specifically, the Paper made numerous references to the international dimension of its drug interdiction efforts, pointing to the commencement of a corollary 'strategic review of international drugs activity' that entailed 'a clear overall commitment of all the law enforcement, intelligence and diplomatic agencies to reduce the flow of illicit drugs to the UK' (1998: 11). Thus, by the time of the 9/11 attacks on the United States, talking tough on drug use and the sources of drug use as a cause of crime was a firmly established policy for the first Blair ministry.

In a keynote speech to the Labour Party Conference in Brighton just five days prior to the commencement of military activity in Afghanistan, Blair laid out his domestic vision of the social harm of drugs within the international context of global terrorism. The speech highlighted the connection between al-Qaeda and the Taliban, and set out providing ancillary, humanitarian reasons why the removal of the latter from power in Afghanistan was the moral and logical course of action, claiming that,

> It is a regime founded on fear and funded on the drugs trade. The biggest drugs hoard in the world is in Afghanistan, controlled by the Taliban. Ninety per cent of the heroin on British streets originates in Afghanistan. The arms the Taliban are buying today are paid for with the lives of young British people buying their drugs on British streets. That is another part of their regime that we should seek to destroy. (Blair 2001b, online)

Blair thereby set out to establish a correlation between the drugs trade, which adversely affected Afghan and European society, and terrorism, which whilst a concern of all western states, was for all practical purposes the United States' sole priority. This statement is consistent with Blair's personal belief in the possibility of eradicating social ills from British (and international) society, and his conviction outlined in earlier chapters of the fundamental interconnectedness, and sameness, of the world's peoples as a result of processes of globalisation. A speech at the Trade Union Congress' Conference in Blackpool reinforced Blair's stern attitude towards drug-related offences, demanding that drug-addicted offenders would be offered an ultimatum of 'treatment' or 'custody' as a means of favouring the victims of crime, the logic being that to protect the innocent, punitive steps must be taken (Blair 2002, online). The idea that coercive measures could rid, or at least ameliorate, the drug problems rife in British society were reflected by the Blair Government's commitments to Afghanistan. The task of rebuilding Afghanistan's internal security was divided into sections amongst the G8 nations at the Tokyo Conference of January 2002: Italy was tasked with reforming the judiciary, Germany was allocated police reform, the United States was in charge of military reform, Japan with reintegration of enemy combatants and, finally, the United Kingdom was given the responsibility of counter-narcotics with the remit 'to rebuild an agricultural system and eliminate poppy cultivation' in the country (Hansard 2002a). Although there is no substantial evidence explaining how and why each state was given each task, it seems a reasonable inference that Blair's penchant for tackling drugs may have played a role in matters. Counter-narcotics was clearly something Blair saw as morally right, strategically useful and, most importantly, complementary to the Americans' counterterrorism efforts. In other words, agreeing to lead nation status on counter-narcotics would allow the United Kingdom to play a major role in the policy end of achieving – through collective effort – the stabilisation and eventual democratisation of Afghanistan. Following the Conference, Secretary for International Development Clare Short reported to the House of Commons the obligations taken on by Government. Short's statement lacked detail and focused only on the UK's responsibility to 'offer people alternative crops' and 'a legitimate life that will be better than the illegitimate life' of producing poppy (Hansard 2002c). Planning for how this would be achieved appeared non-existent. Indeed, Short acknowledged to the House that,

> there is not a firm strategy in place to deal with the poppy crop that has recently been planted and is still in the ground in Afghanistan…we will engage both the Foreign Office and my Department in trying to ensure that such a strategy is in place. (Hansard 2002a)

This statement makes it clear that, from the outset, the narrative of combatting drugs abroad to reduce drug abuse and drug-related crime domestically came before any serious strategic thinking about the subject. In the months following the Tokyo Conference, little explanation was given to explain how a coercive approach to drug interdiction abroad would work. Rather, the narrative suggested a prevailing assumption that, given a legal alternative crop, Afghan farmers would simply reject the production of heroin poppy. Which crop would provide an alternative to heroin, and would be as hardy and valuable, was apparently a theme left unexplored. The substance of Short's address was effectively that all that was required was a firm financial and political commitment from the international community, and the problem could be solved. Meanwhile, the short-term strategy for already-planted poppy crop was one of payoffs. Orchestrated by the Secret Intelligence Service, the 'compensation for eradication' scheme was an attempt to purchase and destroy the entire poppy crop prior to its harvest and processing into opium, on the proviso that Afghan growers would agree not to plant poppy the following season (Meo 2003: 10; Elliott 2001, online; Loyd and Khel 2003: 17). Although the plan managed to destroy roughly one-fifth of the crop, it failed to stem the tide of heroin exports, with much of the allocated funds being siphoned off or diverted into the pockets of power brokers. According to some commentators, the compensation programme had the deleterious effect of actually promoting the production of poppy, with farmers seeing no disincentive since growing the crop would result either in its sale to traffickers or its destruction by eradication teams and eventual recompense (Eagar 2006: 32). In the view of Vanda Felbab-Brown (2009: 106), this represented a moral hazard for the British, for the obvious reason that their compensation scheme enriched middle-men and, by inadvertently incentivising the trade, fuelled the expansion of poppy cultivation. In a narrative sense, it also indicated the relatively simplistic perspective through which British policymakers viewed Afghan society.

This early setback did little to dissuade Government narrative efforts guided directly from Number 10. The counter-narcotics policy for Afghanistan became more strongly linked to domestic drug policy as 2002 went on, particularly in the public statements of senior members of the Labour Government. The phrase 'on our streets' took on special resonance in their messaging activities, with Blair,

Short and Foreign Secretary Jack Straw all commenting, in varying degrees of emotiveness, on the 'mayhem, human suffering and criminality' inflicted on British streets by the Afghan poppy crop (Hansard 2001c, 2002b). Indeed, as was often repeated by such Cabinet members, some 90–95 per cent of all heroin consumed in Britain originated in Afghanistan at this point. The rhetoric of 'on our streets' performed an important role in driving home this point, akin to the counter-terrorism narrative (explored in the next chapter), of bridging the geographical divide between what was happening thousands of miles away from Britain in Afghanistan and in Britain itself. Without the bridging function implicit in such a speech act, Britain's counter-narcotics responsibilities in Afghanistan could easily be construed as having no merit in terms of realist appraisals of national interest.

As with the relating of the counter-terrorism and stabilisation policies to 'national security' or 'national interest', the 'on our streets' rhetoric can be construed as a 'cultural coding' of collective security activity within a national framework of interest and security, utilised to persuade the public of the moral value and strategic relevance of counter-narcotics efforts to British society (Alexander 2011). It also served a narrative function of simplification, by pointing out the direct effect on Britain rather than opting for a more complicated stabilisation story that posited the necessity of counter-narcotics to undermine drug barons and to buttress the new political and economic order. By linking the Afghan opium trade to the social ills of drug addiction in the cities of the United Kingdom, the Blair Government also fulfilled one of its primary philosophical tenets of viewing regional and national issues within a global context. During this period, therefore, an inversion of Labour's transplanting of domestic drug policy onto its international commitments was taking place: by emphasising the importance of Afghan counter-narcotics to UK public health, the counter-narcotic aims of the British in Afghanistan, ostensibly an offshoot of individual priorities within Cabinet reflecting their coercive mentality towards drugs policy in the UK, became a core element of the domestic drug policy.

A preliminary inference is that by attaching foreign policy targets to domestic ones, the counter-narcotics narrative for Afghanistan became fully wedded to the domestic 'war on drugs' narrative. Success or failure in one area could therefore be logically coupled to the success or failure of another. Foreign Office Minister Mike O'Brien summed up the narrative logic at work here, arguing that Britain was 'putting in place policies ... not only because it is right for Afghanistan but also because it will prevent people dying on our streets if we are successful' (BBC 2002, online). These political stakes were raised by Kabul's Afghan National Drug Control Strategy of May 2003, which declared the ambitious aim of *completely eradicating* the cultivation of poppy in Afghanistan within ten years (UNODC 2004a, online). This policy was endorsed by all key players, including the United Kingdom (which committed itself to contributing £70 million over three years) and the United Nations Office on Drugs and Crime (UNODC). Running concurrently with these developments was the publication in February 2003 of the United States' National Drug Control Strategy. This strategy supported

UK-led 'alternative livelihoods' approaches but also called for 'comprehensive eradication efforts', a measure deemed 'politically impossible' in early Foreign Office assessments (Elliott 2001, online; The White House 2003: 38). The American approach, by no means united (due to internal rifts between the Pentagon and the State Department over the correct approach to counter-narcotics), nevertheless at this point viewed drug control within the rubric of counter-terrorism operations, not, as was normatively and institutionally the case with the British, as an end in itself. The White House's overriding concern was disrupting the financing of al-Qaeda, seen largely as generated from opium revenues. The British perspective, meanwhile, appeared more nuanced, acknowledging that the size of the opium industry in Afghanistan effectively meant everyone in the country was directly or indirectly affected by it (Hopkins 2001, online). This difference in priorities led to the beginnings of long-standing friction between the two countries in terms of methods of eradication. Whereas the Afghan strategy and the British approach considered eradication possible only where it would not detrimentally affect stability in the country – that is, where 'alternative livelihoods' could be secured – the American approach increasingly turned toward the need for quicker, non-military eradication techniques, namely through aerial spraying, a technique explicitly disavowed in the Afghan National Drug Control Strategy (Islamic Republic of Afghanistan 2003: 21). The first documented case of aerial spraying in Afghanistan is dated to June 2003, just one month after the release of the Afghan strategy (Harding 2003, online). Despite private opposition by the British to the American approach, any hope of meaningfully pushing back on US demands was complicated by the fact that the resources made available by the British for Afghan efforts were not sufficient to carry out an effective ground-based approach. In early August 2003, the head of the Afghan Counter Narcotics Directorate publicly criticised the British efforts:

> I was expecting Mr Blair to do more. We need funds and assistance. This is not a job that can be done by the Kabul government alone ... My men are dedicated. But they have only tens of thousands of dollars from the UK, not hundreds of thousands. Compare that to spending on the war on terror. (Meo 2003: 10)

The British approach in this first phase of the conflict quickly unravelled. It found itself in the unenviable position of attempting to placate two diametrically opposed parties in the United States and Afghanistan, and was unable to criticise either. The narrative framework set out by the Blair Government was one that championed the necessity – both moral and strategic – of combatting the opium trade in Afghanistan, not just for the well-being of British citizens but also for the internal stability of the Afghan economy and social and political systems. By making this case in narrative, and then following it up by agreeing to take the lead on counter-narcotics at the Tokyo Conference, the Blair Government left itself with little rhetorical or political space in which to negotiate. An idealistic interpretation of the situation on the ground in Afghanistan and the resources required to effect meaningful change contributed to a rapidly deteriorating situation in the country during the first two years of the conflict, indicated by an increase in cultivation of something approximating 1,400 per cent between 2001 and 2003 (UNODC 2002, 2004). The policy, in short, was a failure; but abrogation of responsibilities would have meant not just a party political failure for the Government but also one that would have repercussions in its international standing, as well as opening it up to the potential for attacks from the public that it had 'gone soft' on drugs or from countervailing voices that could deconstruct and lay bare the discrepancies between the original policy, events on the ground, and the policy narrative that accompanied those events. Blair's Government had set itself a 'discourse trap' (Michaels 2013) with what would amount to two equally unsavoury options going forward: continue with a failing policy (and remain constant with the upbeat assessments of the accompanying narrative), or jettison the strategy and risking the political backlash of a defunct policy narrative.

Phase II: 2003–2005 – 'Fitful Progress'

Following the handover of reconstruction and stabilisation responsibilities to NATO/ISAF in August 2003, the counter-narcotics situation in Afghanistan became fraught with inter-state tensions. The American position of treating counter-narcotics as a concern subsidiary to counter-terrorism remained intact in the late summer of 2003, with officials voicing fears that enhanced interdiction measures would alienate their warlord partners and adversely affect their ability to prosecute the Taliban and al-Qaeda; one American official remarked succinctly that 'if you take the warlords out, the whole system of government goes' (Norton-Taylor and Astill 2003, online). The British and the UNODC, meanwhile, attempted to steer US opinion towards a point of view that appreciated the interconnectedness of international terrorism and the narcotics industry. A British official in Kabul argued that the huge sums of money at the disposal of drug 'kingpins' could end up in the hands of 'any number of groups', whilst Antonio Maria Costa of the UNODC made reference to the potential for Afghanistan to become a 'failed state' if 'narco-terrorists' were not dealt with using 'energetic measures' on the ground (Loyd and Khel 2003: 17; Loof 2003, online). The term 'narco-terrorist' was also used by Afghan President Karzai in public addresses (Hansard 2004d).

By early 2004, following a British-convened counter-narcotics conference in Kabul to 'promote greater sharing of the problem', which reiterated the British and Afghan view of the need for a variety of levers and the importance of ground-based eradication, the Americans appeared more open to giving greater political attention to the opium issue (UNODC 2004a, online). However, this piquing of interest came with a series of scathing attacks on British counter-narcotics efforts, culminating in a thinly-veiled statement by Assistant Secretary of State for International Narcotics and Law Enforcement Robert Charles that concluded British efforts were too slow, insufficient in scope, and under-funded (Charles 2004, online). While Charles' testimony placed, for the first time, the focus of counter-narcotic efforts more firmly within the purview of a stabilisation operation rather than just a counter-terrorist one, it also re-emphasised the American view that 'the current set of eradication targeting criteria, while designed with the best of intentions, may be overly restrictive' – essentially meaning that aerial eradication and a reduced emphasis on the centrality of alternative livelihoods programmes were necessary measures.

The timing of Charles' statement, on the eve of the Berlin Conference declaration (which affirmed the need for wider and more intensive participation in counter-narcotics efforts), was conspicuous in its conflict with the Afghan Government's Declaration on Counter-Narcotics issued at the Conference on the same day. In contrast with the American approach, the Declaration provided no mention of aggressive eradication measures, instead choosing to focus on regional cooperation and interdiction exclusively (Government of Germany 2004, online). On the other hand, the main Declaration of the Berlin Conference appeared to give succour to the American approach in stating that Afghanistan and its international partners would 'do everything … to reduce and eliminate' poppy cultivation (IAF 2004: 2). A wider tension between American and Afghan approaches represented something of a circular argument: the Americans proposed that counter-narcotics efforts must be rapidly increased in order to stabilise the Afghan regime, whereas the Afghans themselves believed their regime must be stable beforehand in order for counter-narcotics efforts to be successful (IAF 2004, online). In practical terms this clash of views meant that, for the Americans, the Afghan government could not stabilise whilst the opium economy operated unabated, while the Afghans themselves believed attacking the opium economy too aggressively and without sufficient alternatives in place would only serve to undermine its authority and legitimacy. Opposition Members in the House of Commons appeared to take note of this discordance of views; Conservative Member George Osborne in particular raised the question of whether there was

> a tension between the desire of the international community to create a strong central government in Kabul and the desire of the American-led forces hunting al-Qaeda, who use regional warlords to help them in that important fight. (Hansard 2004a)

The official response of the Labour Government was to downplay any rift between the United States and Afghanistan and its European partners regarding the best means of reducing opium production. Whilst there was some concession by the then-Foreign Office Minister with portfolio for the counter-narcotics effort, Bill Rammell, of 'differences of emphasis' between counter-narcotics contributors, the overarching narrative theme was of unity of effort and coherence going forward. In May 2004, two months after the Berlin Conference, Rammell released a statement to the Commons which reemphasised the British commitment to working in cooperation with, and strategic deference to, the Afghan Government, but also included several references to increased eradication efforts and funding

(Wintour 2004: 12). Rammell's comments represented a detailed statement of British counter-narcotics policy after Berlin, outlining the British position of strategic patience and the imperative of alternative livelihoods. Significantly, Rammell spoke in response to an impassioned speech by Labour backbencher David Cairns, a strong advocate of the counter-narcotics mission; In making the case for strategic patience, Rammell's comparatively measured tone had a considerable element of managing expectation to it, warning Cairns that

> We should have a sense of realism … about the time scales that will be necessary if the problem is successfully to be tackled. Elsewhere in the world, experience of successfully tackling the problem suggests that it will take up to a decade to achieve that in Afghanistan. (Hansard 2004d)

At this time British narrative position was essentially sound in the sense that it fairly accurately matched the scale and scope of the mission at hand. It placed utmost importance on the long-term goals of counter-narcotics strategy, and emphasised the need for patience so that the alternative livelihoods system could take root. In what could be perceived as a message to the Americans as much as any other party, Rammell pressed this issue in his Commons statement, arguing that

> [a]nyone who considers the issue will conclude that if we are to persuade poppy farmers to stop undertaking that activity, they must be given an alternative income. That is why DfID is so engaged, why it has trebled its budget and why this is a particularly important strand of its activities. (Hansard 2004d)

Judging by the public statements coming from the Foreign Office in the months following Rammell's address, considerable progress had been made in synergising UK-US-Afghan counter-narcotics policies. In August, the FCO announced that all parties now were working to a 'common agenda and shared commitment for next year across the whole range of counter-narcotics work' (Smith 2004, online). Despite this affirmation, the autumn of 2004 was met with a series of news articles carrying American critiques, often on-the-record, of the British counter-narcotic approach. Requests made by the UK for the Americans to employ forces to assist in ground-based interdiction and eradication were publicly rebuffed by one official, who admonished the UK for its 'naïve' approach, arguing that going after drug lords would undermine any leverage ISAF and OEF had in the restive Pashtun provinces of the south and east of Afghanistan (Smith 2004, online). The US criticism of what they perceived as a sluggish time-scale cropped up on several occasions during this period as well. Doug Wankel, a Kabul-based State Department official, commented to the press that, from the US perspective, the British

> don't seem to have the same sense of urgency… where we see it's not moving at the … level or the speed we think, we're going to step in and we're going to work with them to help them get it to the level and to the speed which we think it

needs... We really believe that within two years, we've got to see the pendulum
swing. (Zoroya and Leinwand 2004, online)

This statement, similar to several others made at the time, made it clear that the US
were moving to take greater control over the counter-narcotic mission by increasing
operational tempo and shortening time frames for success. The British long-term
strategy and the narrative themes of alternative livelihoods and strategic patience
were simply no longer compatible with what the Americans were saying publicly,
who were still largely of the view that counter-narcotics was, first and foremost,
a counter-terrorism issue rather than a stabilisation issue. The empirical evidence
available suggests that, following these public admonitions by the United States,
the Foreign Office opted to acquiesce on its core themes in the face of American
pressure. Rammell would later tell the press that Government had 'the plans and
the strategy in place to meet our targets and begin to reverse, I would hope, the tide
by this time next year' (Zoroya and Leinwand 2004, online).

In the face of increased critical attention from the Americans, the British policy
narrative positioning had changed considerably in a matter of mere weeks, moving
from a message that emphasised long-termism and a goal of ten years, to one that
pinned long-term success on a tangible reduction in real terms of poppy cultivation
within just one year. Whilst continuing to press for a comprehensive approach
that prioritised alternative livelihoods and intensive ground-based interdiction
and eradication efforts, the narrative approach taken by Whitehall began to reflect
an attempt at diplomatic balancing between the US and Afghan points of view.
Events that followed made this effort progressively more difficult, however. In
the remaining two months of 2004, the US rhetoric on counter-narcotics grew
stronger still, indicated by increasing calls in the United States Congress for aerial
spraying. In November 2004, President Bush validated these calls by approving a
fund of approximately $150 million earmarked for aerial eradication (Jones 2006,
online). Incidentally, by the end of November reports emerged of a 'mysterious'
aircraft engaged in spraying in Nangarhar province in the east of Afghanistan, for
which the United States denied responsibility (Burke 2004, online). The Karzai
government responded to this news with condemnation, whilst local people began
to report illnesses amongst their livestock. Whether or not the correlation between
Bush's funding of aerial spraying and the reported spraying incident were causally
related is unclear, but a blight of Afghan crops (including but not limited to opium
poppy) likely contributed to a drop in cultivated area and opium tonnage in 2005,
representing the first such reduction since 2001 (UNODC 2005: 50).

Despite the widespread outrage at the spraying event, ground-based eradication
efforts during this period also incurred unwanted consequences. Attempts at
eradicating poppy crops were often met with hostility and occasionally with violence
by Afghan farmers, to the point where UNODC monitors were unable to carry out
surveys in several districts (Felbab-Brown 2009: 107; UNODC 2004b: 78). In
Nangarhar province, where counter-narcotics operations achieved an 80 per cent
reduction in cultivation in one planting season, locals complained that promises of

aid and alternative livelihoods were not forthcoming and that the alternative crops had effectively ruined their local economy (Coghlan 2005b, online). The UNODC confirmed this by commenting that it was only 'at times' that aid and development money complemented eradication efforts (UNODC 2005:iii). Other commentators noted that for the goal of complete eradication to be met without destabilising Afghan society and its economy, there would need to be a vast increase in human and financial resources committed to the country (Loyd 2004: 20). Throughout this second phase of the conflict, however, this level of commitment simply did not exist. Multiple impediments to delivery have since been identified, all of which stand in sharp relief to the rather simplistic narrative begun by Blair in late 2001 and illustrate the dangers of creating non-negotiable ethical and moral imperatives in narrative. Firstly, the aid money given was sometimes siphoned off by local governors, who were often implicated in the very trade they were tasked to clamp down on; inequities in land ownership and therefore in political power meant that the farmers too often did not receive their share of the money. Secondly, the scale of infrastructural work, such as field irrigation networks and road networks for transporting goods, was underestimated. Thirdly, the issue of what alternative crops could deliver as much revenue for farmers remained unresolved, with opium bringing approximately ten times the profit one could derive from growing wheat. Fourthly, it is possible that imbalances between 'sticks' and 'carrots' may have resulted in the impoverishment and subsequent alienation of Afghan farmers.

On the one hand, a more coercive approach to counter-narcotics appears to have made an impact in 2005, with over 50 per cent of Afghan farmers polled who no longer grew poppy citing fear of imprisonment and crop eradication as being the main determinant in their decision (UNODC 2005: 60). However, the lack of visibility of available alternatives to poppy cultivation in the face of such coercive measures only served to harden local attitudes towards the counter-narcotic strategy, evidenced by the return to full planting in Nangarhar in late 2005 and the determination amongst some farmers that fresh attempts to eradicate their crops would be met with violent resistance (Coghlan 2005a, online). By the end of 2005, the shortcomings of the British alternative livelihoods effort were fully exposed, with both the Americans and Karzai openly criticising them as woefully insufficient (Karzai 2005, online). Fifthly (and perhaps most importantly) were the issues beyond the immediate control of any national counter-narcotics strategy, namely the weather and the global market. Good weather results in high yields, bad weather in low yields. High yields may mean less acreage cultivated, whilst low yields may mean more land usage. High market prices for opium may stimulate planting, whilst low prices may discourage planting. It would be a mistake, therefore, to assume a basic correlation between success in counter-narcotics and a reduction in cultivation, since this could be indicative of any of these extemporaneous factors. In short, the natural forces guiding the Afghan opium industry had at least as much impact on the opium economy as any concerted state-led efforts, evidenced by cyclical patterns of growth and decline between 2001 and 2005.

All of these obstacles to Blair's original counter-narcotics policy position did not have the effect of giving the Government pause to reconsider either the efficacy of their strategy or the parameters of the narrative employed to explain and defend it. While the British narrative by mid-2005 readily accepted that these complications and contingencies were affecting efforts to meet targets, the overriding message coming from ministers was that the failings incurred simply indicated a need for greater effort in persisting with counter-narcotics activities. The notion that counter-narcotics required a more comprehensive approach that included broad reconstruction, development and governance efforts was cited widely by Government ministers. Because such efforts were seen as necessary to buttress counter-narcotics work, it is vital to note that stabilisation and reconstruction efforts from 2005 onwards were ultimately driven by the counter-narcotics agenda which, of course, had as its own end the shrinking of financial resources available to terrorist organisations. Eradication and alternative livelihoods were ascendant in the policy agenda; counter-terrorism, by contrast, had by this stage of the conflict faded in importance, and many figures within the Afghan and American governments downplayed and even discarded the threat of terrorism and insurgency to Afghan stability (BBC 2004, online). Rather, the focus at this point was almost solely on counter-narcotics as the most important aspect of stabilisation.

This perspective was compounded by a rapid rise in 2005 of insurgent activity as ISAF forces continued their counter-clockwise expansion across Afghanistan; as stabilisation forces came into greater contact with insurgent forces, so too did the perception increase that counter-narcotics and stabilisation were inseparable elements. By late 2005, when British deployment to Helmand province was confirmed, the counter-narcotics agenda was portrayed in media reports as the number one priority in stabilisation efforts. One senior military official told *The Sunday Times* that '[f]rom a military perspective, we see Afghanistan as something that can be resolved but we need to ... realise that counter-narcotics is central', whilst a spokesperson for the Ministry of Defence informed *The Scotsman* that while counterinsurgency activity was necessary, 'breaking up this opium business is just as important. It is a building-block' (Lamb 2005: 2; *The Scotsman* 2005, online). In a speech delivered at NATO headquarters in Brussels, Defence Secretary John Reid declared that 'drugs are now the single greatest threat to Afghanistan's long-term security and stability' (Reid 2005, online). The message coming down from state institutions was unequivocal, and would remain so throughout the first half of 2006: defeating terrorism required the destruction of the opium industry, which could only be achieved by intensive, military-led reconstruction. In an attempt to shore up the discourse trap started by Blair in 2001, the mission in Afghanistan began to expand in scope rapidly, but was oriented around the premise that the success of counter-narcotics efforts was the key to stabilising the country.

Phase III: 2006–2009 – Helmand: Rise and Fall of Counter-Narcotics

As noted in the preceding stabilisation chapters, the balance of available evidence suggests that the policymaking process for Britain's foray into Helmand Province was shaped in a kind of perfect storm of interests between ministers and military officials. There exists little in the way of primary documentary evidence surrounding the decision-making process within Downing Street in the run up to the Herrick deployment in the spring of 2006. Michael Clarke's study of the policy process in the months prior to deployment underlines the point about a lack of paper trail; he commented that 'no evidence has emerged either from documents or interviews that the choice between Helmand and Kandahar was regarded as strategically important by the British' (2011: 16). Of course, it would appear that Clarke's thinking about strategy entailed a traditional view of relating the ways and means of policy and operations with the end of stabilising and/or democratising Afghanistan. Throughout this work, however, I have argued that understanding Britain's actions regarding Afghanistan is indeed challenging if one views its strategy as being undertaken with Afghanistan as the end of policy. Rather, it is useful to conceive of Britain's preparations for, and initial work in Helmand as a consequence of two strategic factors: the first of these being the strategic requirement of achieving unity of effort between the Government and the military, and the second being the strategic requirement of demonstrating Britain's utility and loyalty to their American partners. If one is to conceptualise Britain's policy decisions and strategic direction through this lens, it becomes instantly apparent that all of the apparent blunders of participation in Afghanistan begin to make sense.

In the preceding chapters we have considered the possibility that Helmand was the result of international political pressure from the United States, maintaining the coherence and credibility of NATO, or as a means for the British military to redeem itself from its perceived failure in Basra in the eyes of the martial world by using Afghanistan as 'an opportunity for good soldiering' (Personal Interview 2013). Indeed, in many ways deployment to Helmand fulfilled the counter-terrorism, stabilisation and counter-narcotics narratives simultaneously, even allowing them (perhaps coincidentally) to coalesce together for a brief time. In counter-terrorism terms, Helmand restored the British military to a leading role in the 'good war' (as opposed to the highly controversial and divisive Iraq War) that was more clearly linked to prosecuting the war on terror. In terms of stabilisation, the mission worked because it was marketed to the public as primarily 'non-kinetic' in nature, focusing rather on reconstruction and protection of civilians and the fledgling 'democratic' Afghan state. Finally, with respect to counter-narcotics, the Helmand deployment allowed the United Kingdom to enter into Afghanistan's opium production heartland. With an ostensibly comprehensive package of security, development and governance at its disposal, it offered the Blair Government an opportunity to vindicate its stabilisation-based approach to counter-narcotics and to truly take lead nation status within the collective security framework.

In the months preceding the deployment of 16 Air Assault Brigade, Government ministers – most notably Defence Secretary John Reid – provided public statements laying down the rationale for and purpose of the Helmand mission. In a manner reminiscent of the Blair Ministry's narrative during the first months of the intervention, Reid spent much of late 2005 and early 2006 reiterating the dangers of Afghan heroin 'on our streets', a point made more relevant by the conspicuousness of the phrase's absence between 2002 and 2004 (Hansard 2005b, 2005c). In late January 2006, Reid outlined to the Commons the Helmand mission in detail. He began by identifying three main culprits to be combatted: the Taliban, terrorists, and drug traffickers (Hansard 2006a). International security from terrorism, regional security from the Taliban, and national security from the Afghan drugs trade were identified by Reid as issues of vital importance:

> We cannot risk Afghanistan again becoming a sanctuary for terrorists. ... We cannot ignore the opportunity to bring security to a fragile but vital part of the world, and we cannot go on accepting Afghan opium being the source of 90 per cent. of the heroin that is applied to the veins of the young people of this country. For all those reasons, it is in our interests, as the United Kingdom and as a responsible member of the international community, to act. (Hansard 2006a)

In making the case for viewing each of these issues as in the interest of the United Kingdom, Reid effectively argued that they were of equal importance. Reid further articulated this point in stating that

> we cannot look to resolve just one of those issues. Everything connects. Stability depends on a viable, legitimate economy. That depends on rooting out corruption and finding real alternatives to the harvesting of opium. That means helping Afghanistan to develop judicial systems, her infrastructure and the capability to govern herself effectively, which in turn brings stability and security. (Hansard 2006a)

Since everything connected for Reid, counter-narcotics was clearly an integral part of British planning for operations. The centrality of counter-narcotics in Reid's statements led some to question whether the decision to enter Helmand – as opposed to Kandahar, the spiritual capital of the Pashtun resistance in the south of the country – was primarily motivated by the counter-narcotics agenda. Clarke (2012: 16), for example, argued that personalities played a key role, claiming that the 'Foreign Office was content that UK forces should go to Helmand, and Prime Minister Tony Blair was personally keen on the province – the heart of Afghanistan's narco-economy – as that would be consistent with the anti-narcotics lead role'. Clarke's incredulity over the strategic wisdom over the decision to deploy to Helmand is by no means unique, however; former ISAF Commander and Chief of the Defence Staff General Sir David Richards also went on-the-record, stating that he has 'never yet had a good reason given me why that decision was taken' (Fergusson 2009: 233). This is perhaps the crucial question for both the strategy and narrative for Afghanistan from 2006 onwards: was deployment to Helmand largely a consequence of counter-narcotics policy? According to a report from the House of Commons Defence Committee in April 2006, the Ministry of Defence had said that 'it had chosen to deploy in Helmand Province specifically because it was an area containing continuing threats to stability from the narcotics trade, the Taliban, and other illegally armed groups' (Defence Committee 2006: 16). Indeed, events on the ground prior to the British deployment support the Ministry's statement. Specifically, one may point to the decision to oust the governor of Helmand, Sher Mohamed Akhundzada, at the UK's request due to his links to the opium trade. Akhundzada was replaced by former Afghan counter-narcotics head Mohamed Daud – a clear indication of intent on the part of the British (Jones 2008: 14; King 2010b: 71).

Other factors may have played a role, however. The most convincing of these is the simple fact that the decision to expand ISAF's jurisdiction to the south had been taken, that it therefore required leadership from NATO member states, and that few of those members were willing to volunteer; Britain – in its desire to play a leading role in Afghanistan for all the reasons aforementioned – took on the mantle of responsibility. Within this reading of events, it was posited that part of Britain's rationale for taking on Helmand specifically (as opposed to other provinces in the south) was a result of negotiation with NATO partners, particularly Canada, who wanted to be posted in Kandahar; Britain's desire to give greater roles to NATO partners may have swayed their decision to take Helmand as a concession to the Canadians (Clarke 2012: 15). Of course, it may simply be that what was allocated was largely beyond the control of the UK. Equally, Canada's preference for Kandahar might also have been a consequence of not wanting responsibility for Helmand and the counter-narcotics responsibilities entailed: as one Canadian military officer would later put it, their forces 'have nothing to do with poppy eradication. We stay away from it as far as we can' (CTV 2007, online; PBS 2007, online). Similarly, one may draw an inference from negotiations with the Dutch government, which stalled on its level of commitment to the ISAF expansion to the south, as perhaps informing the allocation of their forces to the more benign Uruzgan province (Baker 2005: 36). A combination of institutional politics within NATO and the personal and institutional enthusiasm on the part of the British for development within a narcotics environment is likely the most accurate portrayal of events.

A further development in the counter-narcotics narrative occurred several days after Reid's Commons statement with the convening of the London Conference on Afghanistan in early February 2006. Serving as the successor template to the Bonn and Tokyo conferences of late 2001 and early 2002, the Conference set out the aims and objectives for state-building in Afghanistan until 2010. Held in London and chaired by Blair (ostensibly as a display of solidarity and leadership), the Conference produced the Afghan Compact, which set out a range of development and security initiatives. Counter-narcotics appeared here as a 'vital and cross-cutting area of work' and set a target of achieving 'a sustained and significant reduction in the production and trafficking of narcotics with a view to complete elimination' (London Conference 2006: 4). Interestingly, although practically all other aspects of the Compact required completion by the end of 2010, counter-narcotics targets remained somewhat indeterminate, with most benchmarks alluding to merely having structures in place by the end of 2010. This represents something of a lowering of expectations, particularly when judged against previous determinations of the Blair Government to turn the tide in opium production or completely eradicate it within a similar timeframe.

By the time Herrick IV began in April 2006, the limitations of the counter-narcotics narrative were almost immediately laid bare. Development and reconstruction opportunities for the British Provincial Reconstruction Team (PRT) were put on hold as 16 Air Assault Brigade found itself in an intensive and protracted kinetic campaign, losing 30 soldiers in the process. The commander of Herrick IV, Brigadier Ed Butler, would later describe conditions on the ground in Helmand as constituting a 'semi-permissive environment', meaning that the ability for the British Task Force to undertake any development work, much less counter-narcotics, was more compromised than they had hoped (Defence Committee 2009:Ev17). However, a larger theme was likely at play here. Perhaps reflective of these conditions or, equally as likely, reflective of institutional tensions over responsibilities foisted upon them by Downing Street and the Foreign Office, the military's appetite for counter-narcotics was never very substantial. Indeed, General Richards would later tell one author that counter-narcotics 'was never on the Army's agenda' (Fergusson 2009: 228). Indeed, this opinion was evidently shared by commanders at brigade level as well; in a media interview at the start of the Herrick mission, Butler acknowledged that counter-narcotics efforts could lead to 'destabilisation' in Helmand (McGrory and Hussein 2006: 31). As 2006 wore on, the compatibility of the themes of reconstruction and counter-narcotics were increasingly called into question in the national press, most notably by Simon Jenkins of *The Guardian*, who questioned the logic of attempting to combine the counter-narcotics and stabilisation briefs into the UK military commitment (Jenkins 2006: 31). In early May, the day before Reid left his position as Defence Secretary to take up a new position as Home Secretary, he responded to a question in the Commons regarding the cooperation between those involved in the opium trade and Taliban insurgents. His response was to the negative, stating that

> [t]here are reports that Taliban have been encouraging Afghan farmers to grow opium poppy and offering protection to farmers against eradication of their poppy crop. I am not however aware of conclusive evidence of a direct link between drug traffickers and the Taliban leadership in Afghanistan, although both benefit from instability. (Hansard 2006d)

This represented nothing short of a complete reversal of the position taken by Reid and all other Labour Government officials in the preceding years. The narrative that had begun on a proposition that the Taliban was a regime 'funded by the drugs trade' had morphed into one that attempted to downplay the linkages between the drugs trade and the Taliban. Rather than admit the incompatibility of counter-narcotics and stabilisation, Reid would instead undermine a fundamental tenet of the counter-narcotics narrative in casting doubt on the centrality of narcotics to the insurgency – that the Taliban, terrorists and drugs were all 'absolutely interlinked'. The narrative line was breaking: it no longer unequivocally argued that 'everything connects'. The relevant point here is not so much that Reid was incorrect in his early assessment; on the contrary, insurgency and narcotics were always linked and remain interconnected to this day. The problem, rather, was that the narrative that initially embraced the connectedness of these issues began to direct the strategic options of the campaign, and by emphasising the linkages between insurgent activity and the narco-economy it became more difficult to engage with either issue, precisely because they were inseparable. This point would be iterated several times over the remainder of 2006 by senior military officials from both the UK and the United States, most prominently by outgoing Chief of the General Staff General Mike Jackson, who referred to counter-narcotics efforts as 'counterproductive' to stabilisation efforts (Fickling 2006, online). Upon Reid's departure from the Ministry of Defence, his successor, Des Browne, took a narrative approach that reflected the military opinion that counter-narcotics was ancillary to stabilisation and potentially compromising of it. In early 2007, Browne repositioned the counter-narcotics policy narrative to a lower priority than under Reid, remarking to the Commons that,

> [i]t is important that we do not allow Afghanistan to become a state that is dependent on narcotics, as too much of its GDP currently is. Narcotics can fund the forces that undermine the Government of Afghanistan and allow it to become a failed state, and have allowed it in the past to become a training ground for terrorists. However, our fundamental objective is to support the democratic Government of Afghanistan and allow their writ to run across the country, so that never again will we, the developed world, be subject to the possibility of terrorist attacks emanating from the failed state of Afghanistan. (Hansard 2007a)

Browne's comment to the Commons represented the beginnings of a deflation of the counter-narcotics narrative in favour of one that prioritised stabilisation. Rather than asserting, as Blair and Reid did before him, that narcotics, insurgency and terrorism are intrinsically interconnected and therefore strategically inseparable, the emphasis here is on the *potential* for narcotics to destabilise Afghanistan and of treating stabilisation and counter-narcotics as possibly distinctive elements. Instead of being a 'vital' element equal to that of stabilisation and counter-terrorism, counter-narcotics was relegated to the status of mere 'importance'. As Blair's tenure wound down in the spring of 2007, the narcotic discourse had altered markedly in the space of just a few months. The 'on our streets' line had all but disappeared from the Government narrative. It had instead been co-opted by Parliamentary critics of the counter-narcotics policy to highlight the failings of the counter-narcotics strategy and Government drug policy in general. Labour backbenchers such as Paul Flynn were quick to use the phrase, for so long utilised by ministers as a sacralising code for framing counter-narcotics efforts as a national issue, as a means of pointing out that the street price of heroin in the UK was cheaper than at any point in 30 years (Hansard 2006c, 2007b). The facts on the ground in Afghanistan were held by critics such as Flynn to be directly responsible for the low cost and high availability of heroin in Britain. By the time of Blair's resignation from Downing Street, the Afghan opium crop was the largest not only since the start of the conflict, but at any point since the Taliban assumed power in 1994. The amount of opium in Helmand province contributed well over half of the record breaking 2007 crop, and cultivation in Helmand itself had increased by 48 per cent compared with 2006, and by 288 per cent compared with 2005 (UNODC 2005, 2006, 2007).

Following the unwelcome news of the 2007 opium crop, familiar themes emerged in the behaviour of Britain's partners. In the United States, members of Congress, the Department of State, and various government departments such as the Drug Enforcement Agency (DEA) led a new push for aerial eradication in Afghanistan (Hemming 2007, online; Semple and Golden 2007, online). As though to underline this view, the Bush Administration appointed William Wood, also known as 'Chemical Bill' for his advocacy of spraying techniques, as their new ambassador to Afghanistan as the 2007 harvest wound down. Intra-departmental tensions remained strong over this period, with the Pentagon and the military brass still highly sceptical of militarising counter-narcotics (*USA Today* 2007, online). Similar rivalries also emerged within the Afghan state: whilst President Karzai remained adamantly opposed to aerial eradication, vice president Ahmed Zia Massoud spoke out publicly in favour of the technique, and in the process severely criticised the British effort in an article for *The Telegraph* (Massoud 2007: 22). Britain's credibility and ownership of the counter-narcotics strategy was severely compromised by mid-2007 when Gordon Brown assumed Prime Ministerial duties. As with the stabilisation narrative, the rhetorical approach taken by Brown to counter-narcotics was decidedly more temperate and analytical than the moralising approach of Blair and Reid. Judging by his public statements, Brown did not share their views on the 'evils' of drugs. In his maiden speech on the subject in December 2007, there was no reference to the link between Afghan opium and British heroin addiction. Indeed, in a characteristically in-depth statement given to the Commons, Brown devoted just over one hundred words to counter-narcotics out of a total address of 2,500 words. Drastic increases in troop casualties and a corresponding upsurge in media attention, coupled with a near-constant assault on Brown's budgeting for the war by the Conservatives and several serving or retired high-ranking military figures all likely contributed to a reduction in the frequency and emotiveness of counter-narcotics rhetoric. Brown's statement also appeared to distance the Government from the 2003 and 2006 benchmarks for poppy cultivation reduction whilst moving the responsibility for such efforts more directly onto the Afghans:

> I am not setting a target. What I will say is that while the combination of the measures that we outlined is necessary, what is also necessary is a central Government who are prepared to take the action. That is why I am impressing on President Karzai the importance of his taking a lead. (Hansard 2007d)

The general policy narrative trend under Brown, therefore, was one of managing expectations and divestment of responsibility, reflecting a loss of control – both in rhetorical and practical terms – of the counter-narcotics strategy. In April 2008 at the NATO Bucharest Summit, member states agreed for the first time to carry out counter-narcotics operations directly (as opposed to the previous arrangement of offering logistical and training support to Afghan counter-narcotics units). The old divisions between civilian and military agencies remained, and Britain's main European partners in the southern region of Afghanistan – the Dutch and the Canadians – continued to display reticence about the wisdom of this increased role *vis-à-vis* their main aim of stabilisation (USA Today 2007, online). Bucharest represented an attempt at a greater 'buy-in' from the international community in counter-narcotics – precisely the kind of formalised burden-sharing arrangement British state had been seeking since the summer of 2006. Whereas the slow drip of authority away from Britain towards the United States since 2006 was replete with public spats and rebukes from the Americans – and resultant about-faces from British ministers – Bucharest signalled this in official and institutionally recognised terms. The perception voiced at Bucharest – that counter-narcotics was a cross-NATO issue – effectively freed the British from the shackles of 'lead nation' status; by November 2008, at a ministerial meeting in Budapest, NATO finally agreed to utilise ISAF forces directly in the counter-narcotics effort (BBC 2008, online). In this sense, the summit represented something of a 'get-out clause' for the discourse trap set by Blair in 2001 and by Reid in 2005–6. The result in narrative terms was an increasing emphasis on 'core' stabilisation objectives in a manner quite distinct from the Blair Government's assessment of the inextricable linkages between narcotics and stability. The same circular argument existed, namely that successful counter-narcotics depends on adequate security, but adequate security depends on the absence of narcotics; the difference at this point was now that it had been agreed that counter-narcotics had been relegated in transnational importance in narrative terms, one could simply focus on the issue of security irrespective of the validity of connections between narcotics and instability. DfID Secretary Douglas Alexander spoke in these terms when he opined that

> although it's a complex challenge dealing with poppies, at its heart is a very simple equation. Where you have law and order and security you can eradicate poppy. And where you have insurgency it's far more difficult. That's why we

have to support not just British soldiers but Afghan soldiers in bringing law and order to those areas affected by the insurgency. (Alexander 2008, online)

Alexander's view that the equation was 'simple' – a linear construct beginning with security and ending with successful counter-narcotics – obviously jars with the original position for Helmand taken by Reid that security, reconstruction and counter-narcotics work must be undertaken simultaneously. Despite such inconsistencies, by mid-2008 the pre-eminence of stabilisation over counter-narcotics was a staple of political communications for the Brown Government. This was only possible because transnational policy decreed it to be the case. There is some question, however, over whether the refocusing on security at the expense of counter-narcotics was a form of political cover for the failure of counter-narcotics itself. To illustrate, Lord Malloch-Brown – then Minister for the Foreign Office – claimed in late 2007 that the 'Taliban do not pose a credible threat to the democratic Afghan government. The Taliban do not control a single province or have the ability to hold territory, showing they are far from being a resurgent force' (Malloch-Brown 2007, online). Seven months later, however, in addressing a Member of Parliament who wanted assurances about Britain's commitments to the counter-narcotics effort, Brown stated that,

[w]e are continuing to fight the war against heroin as well, but ... the reason that we are in Afghanistan is to stop the Taliban taking over there and to stop al-Qaeda coming back in that country. Our aim is to remove the threat to the Afghan people and the whole of Europe, including our own country. (Hansard 2008d)

As much of the counter-narcotics narrative of this phase of the conflict demonstrates, the notion that British forces were in Helmand 'to stop the Taliban taking over there' is a myopic and somewhat misleading portrayal of events. Task Force Helmand's remit was to support Afghan efforts to reconstruct the province and to suppress the drugs trade; it was largely framed as a non-combat mission. Reid's comments in 2006, like Malloch-Brown's in late 2007, do not accord with one that posits a fundamental aim of stopping the Taliban 'taking over'; the narratives they expressed indicated that the Taliban were more of a nuisance than an existential threat to the Afghan state. Of course, such narrative revisionism represents a microcosm of the British counter-narcotic discourse over this period. What started as a, if not *the*, core element of deployment to Helmand, requiring a comprehensive and simultaneous effort, evolved over time to become an ancillary and even counter-productive by-product of stabilisation. Intra-departmental conflict between the military and the civilian bureaucracies in the UK, inter-state tensions over correct methods between the United States, Europe and Afghanistan, and personal convictions over the relative significance of the narcotics trade in Afghanistan all contributed to an almost complete reversal of narrative emphasis between 2006 and 2008. The Brown Government had, through a combination of rhetorical deflation and a loss of control of counter-narcotic strategy, effectively escaped the counter-narcotics discourse trap.

Phase IV: 2009–2011: Abandoning the Policy, Disappearing the Narrative

Obama's accession to the US Presidency in early 2009 signalled a new chapter in allied commitments to Afghanistan. Over a period of several months and multiple strategic review processes, the White House opted to attempt a replication of the population-centric counterinsurgency 'surge' that had ostensibly garnered results in Iraq in 2007–2008. Whilst the American review process carried on into the spring and summer of 2009, Brown faced a series of Afghanistan-related crises, carried out in the full glare of the media, regarding troop levels and alleged equipment shortages. By the end of September, British forces had suffered the deadliest summer fighting season to date, with 57 troop casualties, 22 of which took place in July alone. Amidst the fire-fighting public communication process that ran alongside these unfolding events, there was little to no mention of counter-narcotics measures. In a Commons debate on Afghanistan on 16th July, the Brown Government was criticised for the lack of information on the opium situation. In early August, the Foreign Affairs Select Committee published a report arguing that, in assuming responsibility for counter-narcotics in Afghanistan,

> the UK has taken on a poisoned chalice. There is little evidence to suggest that recent reductions in poppy cultivation are the result of the policies adopted by the UK, other international partners or the Afghan government'… 'the lead international role on counter-narcotics should be transferred away from the UK, and that the Afghan Government should instead be partnered at an international level by the United Nations and ISAF which are better equipped to co-ordinate international efforts. (FASC 2009: 52)

This statement served as a public rebuke of the long-standing narrative line that the UK's counter-narcotics policies were making a difference, clearly contradicting the Labour Government. In testimony published in the Report and received by the Committee in May, Lord Malloch-Brown gave insight as to the state of mind of the Brown Ministry regarding its continued counter-narcotics remit, pointing in the main to alliance dependency issues (and in the process providing a near perfect articulation of the transnational dilemma at work):

> we are trying at least to be a NATO country that meets our share of the responsibility on [Afghanistan]. We are the second biggest troop contributor. We feel that we need to help the Americans by leading on different policy issues where they wish us to. Yes, it is not a comfortable position to be in. It is not great PR to be in charge of counter-narcotics. (FASC 2009:Ev67)

Perhaps fortuitously for the British, the Obama Administration's interest in counter-narcotics proved to be much less than that of Bush's tenure. Obama's appointment to the position of 'Special Envoy for Afghanistan and Pakistan', Richard Holbrooke, wasted little time in criticising eradication efforts as wasteful and counter-productive (BBC 2009, online). In the strategy announcement made by the Administration in March there was an evident shift in direction away from eradication towards more intensive interdiction methods, and the second review address, which incorporated much of incoming ISAF Commander General Stanley McChrystal's assessment, made no mention of counter-narcotics whatsoever. In an ironic turn of events, the reduction of emphasis on counter-narcotics in both the American and British narratives for Afghanistan in 2009 coincided with a considerable reduction in poppy cultivation nationwide (123,000 hectares compared with 157,000 in 2008) and in Helmand (at just under 70,000 hectares, compared with 103,590 in 2008, a reduction of one-third) (UNODC 2009). The reasons for this reduction were quite naturally contested, however. The UNODC's Antonio Maria Costa argued that the improved outlook was the result of counter-narcotic efforts, pointing to enhanced interdiction work and alternative livelihoods programmes. Interestingly, however, Brown and Foreign Secretary David Miliband drew greater attention to the fluctuating commodity prices over 2008 and 2009 as a primary cause for the reduction in poppy cultivation; in 2009, wheat was a far more lucrative crop than opium poppy (Hansard 2009c, 2010a). Their brevity in accounting for market factors beyond the control of a national counter-narcotics strategy may be a reflection of the relative divestment of responsibility of the UK from that strategy, whereas Costa's reasoning could be understood as indicative of his organisation's explicit interest in furthering that strategy. Freed from the confines of the counter-narcotics discourse trap, the narrative coming from the Brown Ministry appeared more transparent than in years previous. The crucial point informing this relinquishment, of course, was that it was only when the United States largely omitted counter-narcotics from its strategic requirements for Afghanistan that the United Kingdom could reconfigure its own policy narrative, even though its own interest in the matter had demonstrably diminished years earlier. This represents clear testament to the power of transnational dilemma in shaping British foreign policy.

The slow divestment of the British role in Afghan counter-narcotics was all but completed following the London Conference on Afghanistan in January 2010. The Conference focused on increasing Afghan ownership of ISAF activities – referred to in bureaucratic parlance as 'Afghanisation' – in correlation with a decrease in direct involvement from the international community (Mancini 2011: 7). In a Commons session in February, Miliband responded to a question by a fellow Member regarding the future of the British role in counter-narcotics:

> the concept has not worked well. It was not formally buried at the meeting last week, but the emphasis that has been placed over the past two or three years on Afghan leadership and international support reflects a recognition of a different, and better, way of doing things. (Hansard 2010b)

In the months between the London Conference and the May General Election, counter-narcotics ceased to exist in narrative terms for the Brown Government. As far as they were concerned, British responsibilities in counter-narcotics were over. Indeed, during the televised election debates preceding the election, counter-narcotics was not mentioned at all and, upon the electoral defeat of the Labour Party and the formation of a Conservative-led Coalition Government, this trend continued. The Cameron Ministry's overall Afghan narrative focused solely on matters of 'national interest', paring down the stabilisation narrative from one that balanced British security needs with a comprehensive view of Afghan stability, to one that prioritised British security and, in doing so, delimited the stabilisation requirements of Afghanistan to the building up of its nascent security apparatus. According to one source, the view within the new Cabinet was that pursuit of counter-narcotics was dead in the water, particularly in terms of the overarching Afghan narrative:

> By the time Cameron had made his timeline, counter-narcotics had got virtually bugger all to do with it. We hadn't made much progress on counter-narcotics, there wasn't much prospect we were going to, it was quite a hard sell to the public that the purpose of Afghanistan was to keep drugs of the streets, and if it was it wasn't working at all. So it wasn't a very appealing part of the narrative... in all honesty it had almost become an irrelevance by that point... It certainly wasn't one of the things I would go into a TV studio and say why we're there. (Personal Interview 2013)

The previous chapter explains how the Coalition narrative agenda was one of simplification, impelled in part by an increased emphasis placed on SC. This informed a desire to keep the stabilisation narrative straightforward and uncomplicated by the confounding logic of counter-narcotics and the comprehensive approach. The consistent failure of British counter-narcotics did not help matters, either. Of course, the counter-narcotics narrative was not a Conservative or Liberal Democrat creation; since the Coalition did not own the narrative, they did not need to defend its historical record. In a speech in September 2010, Defence Secretary Liam Fox provided a succinct verdict of Britain's experiences in Helmand, arguing that

> we assessed that Afghan-grown narcotics posed a greater threat than Afghan-based terrorism. On the basis of the sporadic attacks from Al-Qaeda and their limited effect, it looked as if the threat was being contained. Well, we were wrong, and our presence in Afghanistan now is a consequence of this misjudgement. (Fox 2010b, online)

Fox's speech served as an epilogue for British involvement in counter-narcotics, essentially blaming it for the shortcomings of the mission itself. It confirmed that entry into Helmand was primarily about counter-narcotics, and that the UK's lead nation status on counter-narcotics shaped the course of the mission. Furthermore, it implied that the Government's primary foreign policy responsibility of extricating British forces from Afghanistan was in part a response to the strategic inflation of the original counter-terrorism mission into a counter-narcotics mission, and from that into a nebulous and confused plan to restructure the entirety of Afghani politics and economics. The Coalition mantra that 'we are in Afghanistan for our own national security' can be seen in part, therefore, as a narrative corrective for the rhetorical excesses of previous Governments.

Phase V: 2011–2014 – The Disappeared Policy Narrative

Given the line drawn – both in narrative and institutional terms – under counter-narcotics by the Brown and Coalition Governments, the final phase of Britain's military involvement in Afghanistan contained little regarding the subject. Financial commitments to alternative livelihoods schemes remained in place (albeit at a much reduced amount compared with sums allocated under the Brown Government), but in rhetorical terms counter-narcotics was given only passing mention in any policy statement delivered by members of Cabinet. Indeed, in December 2013, Prime Minister Cameron stated in an interview with reporters that the mission in Afghanistan was for all intents and purposes accomplished, emphasising once more that 'our aim has always been, can we make sure that this country is no longer a haven for terror and terrorists' (YouTube 2013, online). In declaring that the Afghanistan was a 'mission accomplished', however, Cameron made no reference to the tenuous and deteriorating narcotics situation in the country, or Britain's historic role in attempting to 'completely eradicate' the opium trade by the end of 2013.

In the wake of the NATO/ISAF decision to absolve itself of counter-narcotic duties, Russian attention to counter-narcotics in Afghanistan increased, spurred on in part by an influx of Afghan heroin into the country and a corresponding surge in cases of heroin addiction. The Russian response to this domestic crisis culminated in the appointment of former Russian foreign minister Yury Fedotov to the role of Executive Director of the UNODC in mid-2010, which by 2011 was the only foreign organisation that remained directly committed to tackling the issue. Familiar issues dogged their efforts, however, most notably in the guise of fluctuating market prices for the drug, which soared to $241 per kilo for dried opium in 2011, representing an 81 per cent increase on 2010 prices (UNODC 2011). Despite considerable increases in eradication efforts and the continuation of alternative livelihoods schemes funded by the international community, cultivation is currently once more on an upward trend: from 131,000 hectares in 2011, the most recent estimate (for 2013) stands at 209,000 hectares, representing the greatest land use for opium cultivation ever recorded, surpassing even the heights of 2007 (UNODC 2013). In a most ironic turn of events, it was the Taliban during this period that made the headlines for taking measures to eradicate poppy cultivation in the country, representing a reversal in their policy following the lifting of their ban on cultivation in 2001. A spokesperson for the UNODC,

Jean-Luc Lemahieu, heaped praise on their efforts, declaring the UNODC 'welcome this new approach and would hope that this is not a one-time exception but that the Taliban, and others alike, would take a principled stance against the narcotics business' (Graham-Harrison 2012, online). Bizarrely, the international community's counter-narcotics narrative had come full circle: from targeting the Taliban as a 'regime funded by the drugs trade' – with that drugs trade financing terrorism – to one that publicly thanked those who colluded with the terrorists for taking steps to reduce it, quite irrespective of the methods they employed in doing so.

Conclusion

One can hardly disagree with the Foreign Affairs Select Committee's appraisal of the United Kingdom's role as lead nation on counter-narcotics as anything other than a 'poisoned chalice'. What makes the counter-narcotics narrative distinct from the counter-terrorism and stabilisation narratives, however, is the fact that it was largely a poisoned chalice of the United Kingdom's own making, driven as it was by a narrative informed, in the main, by Blair's personal views on drugs and his party's desire to blend American counter-terrorism with British humanitarianism. Helmand in turn can be seen as a redoubling of Labour's preoccupation with the drugs trade, impelled by the UK's desire to placate the Americans after Basra, as well as its need to be seen as a responsible leading nation within the international community and the NATO alliance. Until the counter-narcotics narrative came face-to-face with the difficult realities of implementation in Helmand in mid-2006, it was characterised by a narrative function that sought to link the social ills of heroin addiction in the streets of the United Kingdom to the successes or failures of Afghan eradication, interdiction and alternative livelihoods programmes.

Whilst this was a powerful discursive move, it also produced immense expectation on both domestic and foreign drug policy, where success or failure in one area indicated the same in the other. The pre-eminence of counter-narcotics within the meta-narrative for Afghanistan – framing it as equal to or above counter-terrorism and stabilisation, and as a national security issue in its own right – produced a narrative that propelled policy along with it, effectively locking the UK into pursuing a counter-narcotics agenda that rarely appeared to be working and was always affected by factors – most importantly fluctuations in agricultural commodity prices – that were beyond the scope of a national drug control programme. It was only when those responsible for the creation of the Afghan narrative departed from power – most notably Blair and Reid – that counter-narcotics could be reduced in its relative importance and eventually phased out altogether. Circumstances in other areas, such as an upsurge in insurgent violence and a corresponding rise in troop casualties between 2006 and 2009, and the decision of the Obama Administration to reduce its own emphasis on the importance of counter-narcotics work allowed public officials to ease counter-narcotics off of the policy agenda and out of the Afghan narrative, but only to the extent that they were compelled to admit that the policy had failed, and that the wider Afghan strategy had been compromised by a fundamental misreading of what Britain could achieve with the resources at its disposal. Here, more than in any other case in this work, the creation of a narrative contributed to the relegation of strategic choice below the demands of rhetorical viability – hence the repetition of the stabilisation and counter-terrorism narrative that asserts, quite selectively, that the mission in Helmand was always consistent, and that that mission was not primarily – or even partially – undertaken in order to control and reduce the narcotics trade. Of course, what this chapter demonstrates is precisely the opposite: that it was the counter-narcotics narrative, not counter-terrorism, that informed narrative British explanations of the purpose of its expedition into Helmand specifically; the discourse created its own trap.

Relationships of collective security helped create the counter-narcotics policy and also played a pivotal role in its disappearance as a policy narrative. The Blair Government took on a leadership role for Afghan counter-narcotics because its transnational outlook viewed such work to be in the interest of British society, but conceived of British society as just one component in a wider international social order. It was, as such, the perfect articulation of Blair's dialectical worldview where issues of local and global, national and transnational, military and non-military, war and crime, and social and security could be combined and reconciled as inseparable unities. In this sense, counter-narcotics was, on paper, an ideal means by which Blair's personal philosophy and affinity for a British role in an 'international society' could be fulfilled. However, it is for these same reasons that counter-narcotics policy ultimately came undone: it was always unlikely to succeed because it was never truly 'strategic' for the UK – it did not sufficiently link policy to operations; rather, it was from the first simply policy masquerading as strategy. It was devised by politicians and emerged from a planning process wherein ministers and military officials sought to implement a policy that would serve their respective interests. Rather than developing as a 'strategy', it came about as a consequence of a kind of gentlemen's agreement between civilian and military leadership: the military would be given a fresh start – a new 'opportunity for good soldiering' in the wake of Iraq – and in return ministers would be given an opportunity to prove Britain's significance to the Americans by taking on a leadership role that would also provide an opportunity to showcase its 'comprehensive approach' to stabilisation. This goes some way in explaining the power of the 'policy narrative' in this instance: *because the policy was the strategy, the policy narrative trumped operational reality, even to the extent that the unworkable nature of those operations was misrepresented in order to present a picture that suggested they were, in fact, largely successful.*

Aside from the obvious problems of applying a policy instead of a strategy to an issue, counter-narcotics was dropped because it existed for the purposes of maintaining transnational cohesion and ended up creating discord. On the one hand, it served a purpose of demonstrating British commitment to the Americans as an assumed prerequisite to stabilising Afghanistan. When it became clear that the Americans no longer considered the British contribution to counter-narcotics to be productive – to the point where American officials were issuing hostile public statements to that effect – it became more than evident that counter-narcotics' purpose as a transnationalised policy had backfired. On the other, it served to bind together the Defence establishment with civilian departments (Downing Street, the Foreign Office, and the Department for International Development) by providing a context around which the comprehensive approach could unify effort and maximise output. The lack of strategic consideration in pursuit of such an end inevitably produced tension between planners and operators, however, and with the emergence of Defence-driven strategic communication processes, the policy would be dropped – in similar circumstances to aspects of stabilisation – as it was evidently deemed surplus to defence requirements. It was not a 'national security' priority.

Chapter 6
The Counter-terrorism Narrative

In the preceding three chapters, I have argued that (1) the transnational dilemma has impelled the British state to alter its policy narratives explaining its presence in Afghanistan from one that was based on liberal-centric norms of collective security to one focused on a realist-centric national context; that (2) both of these two approaches to policy narratives created political pressures on the state's ability to articulate its purpose in Afghanistan; that (3) both of these approaches produced a kind of disconnect between what was said about the policy and the requirements of strategy on the ground; (4) that strategic communication practices and processes have sought to avoid the communicative and strategic difficulties innate to the transnationalised policy at the heart of Britain's collective security membership in Afghanistan by working to give communication a greater role in shaping Britain's strategic direction; and (5) that narratives organised around principles agreeable to the Defence community have allowed the British state to partially heal the rifts between Government and Defence.

In this final empirical chapter, the implications of the failings of stabilisation and counter-narcotics efforts in Afghanistan – and, by proxy, the shortcomings of the institutional and normative bases by which those policies were conceived and undertaken – are explored by reference to an analysis of the (chronologically speaking) first and last policy narrative of counter-terrorism. This narrative is now the dominant way British officials account for the history of the Afghan conflict and the performance of the British therein. An example of this can be found in a press briefing given by Prime Minister David Cameron at the UK's Helmand headquarters, Camp Bastion, in December 2013. Cameron used the occasion of his visit to emphasise the beginning of the end of British combat operations in Afghanistan by stating that, in his view, the UK's mission was essentially accomplished. During this briefing he argued that,

> the absolute driving part of the mission is a basic level of security so [Afghanistan] doesn't become a haven for terror. That is the mission, that was the mission and I think we will have accomplished that mission and so our troops can be very proud of what they have done. (BBC 2013, online)

Cameron's statement made the claim that the United Kingdom's narrative for Afghanistan had always been consistent, had always been about countering international terrorism, and that the UK has achieved that objective. Whether or not Britain and its NATO partners have accomplished the mission of securing the country so that it does not once more become 'a haven for terror' is beyond the scope of this work, and is a question that will only be answered in the fullness of time. It should be clear, however, that the issue of exactly what the purpose and objectives of the Afghan mission were in 2013, and what the purpose and objectives were in the years prior to Cameron's statement, are highly contestable, however. The assertion that Britain's mission was successful is only possible if one takes it as given that the primary aim of the intervention throughout its duration was the prevention of further terrorist plots forged in the country. Certainly, the counter-terrorist narrative strand had been present in the overarching Afghan narrative since 2001, and issues of security – whether national (for the United Kingdom), regional (for Afghanistan and its neighbours) or global – have indeed provided the core rationale for the commitment of forces in the country. In an interview with the author, a UK official opined that counter-terrorism has always been at the core of the British narrative for Afghanistan (Personal Interview 2013).

Indeed, such assertions form the premise of much of the analysis that follows. They raise critical questions of whether counter-terrorism has been consistently stated as the reason for British participation, and if other themes existed. Clearly, the preceding two chapters have affirmed the second question. This chapter addresses the first by arguing that the counter-terrorist policy narrative was central to the UK narrative for Afghanistan, but with a number of significant caveats that essentially undermined the consistency of messaging. Firstly, there is the conceptually simple case of counter-terrorism being subsumed by consequent narratives of stabilisation and counter-narcotics. This divergence of narrative away from the core message of counter-terrorism was most pronounced prior to and during the UK Armed Forces' entry into Helmand in 2006, quite understandably as a reflection of the fact that the mission was explicitly framed as not one of counter-terrorism but to do with counter-narcotics and reconstruction. As the mission in Helmand encountered an unforeseen depth of strategic and political obstacles to its realisation, however, the narrative of counter-terrorism eventually and gradually resurfaced, and as a result simplified and streamlined multiple and often discordant explanations of British involvement into one key strand. SC's primary function here was to reinstate counter-terrorism as a central position, unburdened by talk of ancillary rationales. This return informs the second point of this chapter, which explores the variation between early and latter stages of the conflict in terms of the referent object of the counter-terrorism narrative. For Blair, this was not the physical security of Britain but the physical security of *all countries*: the very idea of 'national' interests and security was largely anathema to Blair's internationalist discourse. He rhetorically conceived of security not solely or even primarily in physical terms, but in ideational terms. As we shall see, this understanding of counter-terrorism fit with the prevailing realities of Blair's approach to the conflict. Conversely, the revamped counter-terrorism narrative promoted first by the Brown Ministry and then by Cameron argued, often in hypothetical terms, that Afghanistan was integral to protecting the physical security of the UK specifically, rather than the physical and/or ideational security of the international community as a whole. We can understand this shift in emphasis in part by taking account of the rise and nominal decline of the Blair Ministry's affinity for a vision of the national interest as located within a construct of 'international society', and juxtaposing this in relation to the relative realpolitik-infused messaging of the Brown and Cameron Governments, which favoured a more distinctively nation-centric rhetorical approach and reflected a gradual change in emphasis towards counter-terrorism by the Americans under Barack Obama.

Phase I: 2001–2003 – Counter-terrorism, Internationalism and Humanitarianism

At the heart of the Blair Ministry's response to the terrorist attacks of September 11, 2001 was an appeal – ostensibly aimed foremost at the Americans – to wed humanitarianism to their military activity in Afghanistan. Underpinning this approach was New Labour's 'ethical foreign policy', which promoted a vision of the universality of their normative principles of freedom and democracy. Blair's personal conviction was that the 'War on Terror' could be more than simply a punitive campaign against terrorist transgressors; it could also be (indeed, it must be) a rehabilitative process by linking military activities to developmental activities. In early October 2001, Blair spoke at the Labour Party Conference in Brighton and outlined his vision for a synthesised counter-terrorism/humanitarianism approach:

> out of the shadow of this evil, should emerge lasting good: destruction of the machinery of terrorism wherever it is found; hope amongst all nations of a new beginning where we seek to resolve differences in a calm and ordered way; greater understanding between nations and between faiths; and above all justice and prosperity for the poor and dispossessed, so that people everywhere can see the chance of a better future through the hard work and creative power of the free citizen, not the violence and savagery of the fanatic. (Blair 2001b, online)

In setting out counter-terrorism as complementary with, rather than anathema to the various peace-building and humanitarian considerations at the core of Labour's foreign policy, Blair sought to utilise a narrative genre (Smith 2005; Alexander 2011) that emphasised the binary opposition of the universality of Western values and the apocalyptic threat of Islamist terrorism in order to sway American and world opinion towards his own objectives which, incidentally, as his political philosophy informs us, were in his view really everyone's objectives. As discussed in the previous chapters, Blair and his Cabinet consciously chose to view the national interest of the United Kingdom as inseparable from the collective interests of the international community. Labour's ethical dimension to foreign policy was structured around this principle, which was, in turn, informed by a liberal reading of international relations that held interdependence to have broken down the demarcations between national and global interests. In order to do so, however, a natural consequence was the breaking down of national relationships between domestic interests – be they private, public, or communal – and the foreign policy of the state. In a paper by then-Foreign minister Peter Hain, this principle was articulated as 'convergent policy-making', which posited that 'there is little incentive ... to focus on shared long-term interests in which we shall all lose unless we are willing to look beyond the demands of today's powerful lobbies at home' (Hain 2001: 31). This internationalist streak duly informed the moral substance of Blair's philosophy for interventionism which held, at its core, the precept of acting in accordance with a Rawlsian original position, where all people are imagined as of equal importance when considerations for intervention are made:

> we are a community of people, whose self-interest and mutual interest at crucial points merge, and that it is through a sense of justice that community is born and nurtured. And what does this concept of justice consist of? Fairness, people all of equal worth, of course. But also reason and tolerance. Justice has no favourites; not amongst nations, peoples or faiths. (Blair 2001b, online)

Such positions allowed Blair to articulate the normative and strategic position of the United Kingdom as one in line with the mutual aid principle outlined in Article V of the NATO Charter. Therewith, he could quite reasonably assert that 'what erupted on the streets of New York on September 11 was not an attack on America alone. It was an attack on us all' (Blair 2002a, online). Bearing in mind the ethical dimension prevalent in his post-9/11 rhetoric, it is crucial we recognise that Blair did not simply refer to a physical threat. Blair's first point in his first address to the Commons three days after the terrorist attacks on the United States emphasised that the fundamental threat was to the values of western democracy:

> these attacks were not just attacks upon people and buildings; nor even merely upon the United States of America; these were attacks on the basic democratic values in which we all believe so passionately and on the civilised world. (Hansard 2001a)

In Blair's view, security went beyond material and national boundaries; what was attacked, and what needed to be secured, was also in the realm of the metaphysical and ideational. The concepts of freedom, democracy, human rights and the rule of law were the long-term objects of security and means by which – and only by which – a just and therefore sustainable international security order could be realised. By securing and adhering to this principle for international community, alliances could be bound together under a common normative denominator. As common foundational interests and principles, these values provided the basis for what Blair hoped would amount to a new multilateralism. Simply put, for Blair the 9/11 attacks were 'upon us all' not just, or even primarily, because they indicated an impending threat to citizens and property in the United Kingdom or in mainland Europe or elsewhere, but because they were attacks upon a shared sense of collective morality and well-being. Because of this understanding, security, like interest, could not be conceived of as nationally-oriented under the Blair Government, particularly when related to Afghanistan; 'solidarity' between nations was recognised as the 'route to practical survival' for all states within the international community (Blair 2002, online).

This strand of multilateralism was not necessarily recognised by the Bush Administration, however, which preferred to undertake military ground operations in 2001 almost entirely unilaterally, with the British acting in a naval and air support capacity only. In contrast with Blair's early speeches, the involvement of British ground troops in counter-terrorism activities during this phase was limited to a series of engagements by the Special Air Service and Royal Marines in Paktia and Khost province along the Afghan-Pakistan border in the spring of 2002. Despite the failure of Coalition forces to capture or kill Osama bin Laden, the missions were considered a strategic success and the British battle contingent was drawn down and replaced with a smaller force, based in Kabul. In June 2002, the Afghan Interim Authority held a *Loya Jirga*, or 'grand assembly', of regional leaders at which the constitution for the Islamic Republic was agreed upon by tribal representatives, thereby bringing to a close the initial period of combat operations in the country. Throughout this period, references to counter-terrorism by members of the Government naturally became less frequent as a reflection of changed priorities away from counter-terrorist operations and towards stabilisation and reconstruction. This narrative alteration was reflected in the conclusion of *Operation Veritas* (the British contribution to American-led counter-terrorism operations) and the establishment of ISAF, for which the United Kingdom was the initial lead nation. What references did exist remained true to the original narrative positions of values, internationalism, and multilateralism. In an interview with David Frost in March 2002, Foreign Secretary Jack Straw outlined the continued presence of British forces in the country along these lines, stating that

> the key factor determining our deployment of troops is perfectly obvious, it's the damage that was done on September the 11th and the potential that the Al Qaieda terrorists pose and continue to pose to the security of Afghanistan, to the region and to the world. (Straw 2002, online)

It is worth noting that, throughout this period, the author could not locate a single explicit reference made by any member of the Blair Government that operations in Afghanistan were being performed specifically for the purposes of British national security. As with Straw's statement, counter-terrorism in Afghanistan was related either to that country's stability, the security of the south Asian region, or global security; equating the mission with security in the United Kingdom simply was not a narrative device employed by the Government. This is not particularly surprising, for three rather obvious reasons. Firstly, the purpose of the intervention was, as Straw declared, at this point self-evident, given the relative freshness of 9/11 in the minds of the British public. Secondly, the UK's involvement in counter-terrorism operations was limited in comparison with the Americans. Thirdly, the logic of

Blair's liberal internationalism precluded talk of a distinctively national security policy, in rhetorical terms at the least. A fourth reason relates to the emergence in mid-2002 of Iraq as the foreign policy priority of the United States and, consequently, of the Blair Government. By September 2002, the ratcheting up of rhetoric against the Saddam Hussein regime reached new heights with Blair's announcement in the House of Commons of the 'WMD dossier'; political and media attention by this point was firmly on the controversial subject of invading Iraq, while Afghanistan scarcely featured. When mentioned by members of the Cabinet, the trend was to refer to the conflict element of British involvement in Afghanistan as completed, and to reiterate commitments to the reconstruction of the country. For example, the Defence Secretary, Geoff Hoon, spoke in the Commons of progress made and of drawing down troops in the country as early as October 2002, indicating that the view of Government was that military action in Afghanistan was finished and that, as such, the threat posed to global security by Afghanistan was much reduced (Hansard 2002d). The rhetoric of security by this point had moved onto Iraq, and would remain there until preparations for the Helmand deployment were made public in 2005. Interestingly, the one clear mention of a national security threat to the United Kingdom during this first period related not to Afghanistan, but Iraq, in the form of the potential for states with WMD capability to combine with Islamist terrorist organisations. In setting out the case for intervention in Iraq, Blair drew upon the threat to British security, stating that

> they share one basic common view: they detest the freedom, democracy and tolerance that are the hallmarks of our way of life. At the moment, I accept fully that the association between the two is loose – but it is hardening. The possibility of the two coming together – of terrorist groups in possession of weapons of mass destruction or even of a so-called dirty radiological bomb – is now, in my judgment, a real and present danger to Britain and its national security. (Hansard 2003a)

History has since dismissed the credibility of these claims, and Blair's own belief in their veracity is still open for questioning. The point worth making here is that Blair utilised a quasi-apocalyptic interpretation in order to press the point of the importance of preventing such a possibility – a point which was by no means conclusive, self-evident, or overwhelmingly accepted by parliamentarians or the public (Smith 2005). Crucially, given the high stakes of securing legislative backing for the invasion of Iraq, Blair appears to have grounded his claims not in the internationalist realm of values or in the collective realm of security, but rather in a nation-centric cultural coding of security, ostensibly in order to impress the importance of participating in a potential invasion of Iraq to a sceptical audience. As controversial as the issues of British national security and the WMD dossier were, this was also just one of several reasons Blair and his Cabinet colleagues gave to justify the invasion of Iraq. Most of their rationales remained normatively-based: removing the brutal dictatorship of Saddam Hussein from power and bringing democratic government to the Iraqi people, strengthening international institutions such as the United Nations, developing the transatlantic relationship between Europe and the United States, and the shaping of American foreign policy towards a more multilateral outlook were all cited by Blair as primary or secondary rationales (Hansard 2003a). These were less controversial than the national security claim; in response to that particular rhetorical route, allegations arose that the Government had 'sexed up' the dossier to make the national security issue appear more urgent and existential, typified by the claim that Iraqi WMDs could be deployed against the UK within 45 minutes. With the benefit of hindsight it would seem that, in the face of a 'hard sell' to a sceptical British public unconvinced by humanitarian or internationalist discourses, the Blair Government made the decision to emphasise the imminent risk posed by Hussein to the lives of British people. As we shall see, this theme recurs in relation to Afghanistan in the narrative reset of 2008.

Phase II: 2003–2005 – The Peak of Collective Security Discourse

Given Britain's lead role in counter-narcotics, NATO's takeover of ISAF, the war in Iraq and the relative calm of Afghanistan between 2003 and 2005, the counter-terrorism narrative did not feature much in the speeches and statements of Government officials during this phase of the conflict. Where it did, it was characterised by a totalising view of Afghanistan as one front within a wider War on Terror, where Iraq and Afghanistan were spoken of as being two parts of a coherent whole. In rhetorical terms, at least, it was natural that this was so, since speaking of a *global* War on Terror as the problem – rather than two distinct campaigns with drastically different social, economic and geographic makeups and animated by quite different rationales – fit into Blair's (and Bush's) Kantian meta-narrative for global democratic transformation as the catch-all solution to terrorism and failing states. By claiming both conflicts to be central to and indivisible from a global, existential struggle against terrorism, the decision to pursue two conflicts simultaneously could be given greater authoritativeness. Blair's understanding of the nature of terrorism and the United Kingdom's role in combatting it would not allow any such distinction to be made:

> The truth is that if we withdrew from Iraq, we would be told to withdraw from Afghanistan, and if we withdrew from Afghanistan, we would be told to withdraw from the whole of the middle east, and then we would be made to withdraw even more... the hon. Gentleman asked whether we had not made ourselves a bigger target by our action in Iraq. My answer is that we are a target for these people by our very existence and the values we believe in, and that the only way to defeat them is to get after them. (Hansard 2004b)

Thus, in Blair's mind, terrorism was a security issue for all people and all states, because terrorists do not pick their targets on the basis of what those targets' policies consist of or what they are understood to have done, but because of the values that they represent. Here we can observe a clear link between Blair's value system and unequivocal rhetorical support for collective security operations. Arguing for an understanding of warfare (and of the motives of one's enemy) as one centred upon values rather than tangible material or political factors meant that the Blair Government did not need to perceive the conflicts as separate; rather, because of the logic of interdependence and globalisation that pervaded New Labour's foreign policy, all things and all peoples were connected, willingly or not, in the struggle against international terrorism. Thus, the strategic details and operational implications of foreign policy decisions appear to have been almost an irrelevance for Blair. Since values were the animus for terrorist activity, not intervention and invasion, Britain and all other liberal democratic states were at risk whether they participated in Iraq and Afghanistan or not:

> what now happens in Iraq and Afghanistan affects us here as it does every nation, supportive or not of the actions we have taken. (Hansard 2004e)

In the face of the logic of a values-centric conflict which implied that terrorists would not distinguish between the boundaries of troop contributing and non-troop contributing states (but rather focused on the type of polity each state adhered to), the obligation for multilateralism and interventionism can only appear as a rational imperative. It is worth considering this view on its own merits: despite the difficulties associated with Iraq in mid-2004, notably in the form of a tactical defeat in Fallujah and the breaking of the Abu Ghraib prison abuse scandal, there were grounds for optimism in Afghanistan that supported Blair's internationalist language. The Afghans had ratified a new constitution, the international community had pledged some £8 billion to reconstruction and development efforts at the April Berlin Conference, and the remit of NATO's ISAF mission had been expanded to take eventual control over the security of the entirety of Afghan territory at the Istanbul summit in June. Stabilisation in Afghanistan offered the more reticent European NATO states the opportunity to participate in the War on Terror in a largely combat-free role and, in doing so, validated Blair's counter-terrorism narrative as a truly international and comprehensive effort. In this vein, Blair could reinforce his core points – that the War on Terror was a global, democratic and multilateral response to terrorism, and it was so because terrorism represented a threat to all free peoples and manifested itself most frequently in places around the world where freedom was lacking:

> we will stick with this and see it through, because it is in the interests not only of security in those countries [Iraq and Afghanistan] but of global security. (Hansard 2004f)

To press the point, the Blair Government's focus was on looking at terrorism as a global issue requiring global action for global solutions; there was scarcely a single reference in the Government's counter-terrorism narrative of international terrorism as an explicit threat to the 'national' security of the United Kingdom. Blair's focus was on international and transnational solidarity and the importance of maintaining collective security alliances. Indeed, this theme was echoed by senior NATO officials throughout 2003 and 2004, suggesting that collective security and not counter-terrorism was the animating feature of expanding and consolidating the international presence in Afghanistan. On the day NATO assumed control of ISAF, then-Deputy Secretary General Alessandro Minuto-Rizzo remarked that,

> [t]his new mission is a reflection of NATO's ongoing transformation, and resolve, to meet the security challenges of the 21st century. But most of all, NATO's increased involvement demonstrates our nations' continuing, long-term commitment to stability and security for the Afghan people. (Rizzo 2003, online)

It is noteworthy that Rizzo combines these two elements – NATO transformation and stabilisation of Afghanistan – as the main features of the ISAF command decision. Intriguingly, while he states that the stabilisation imperative was the 'most' important of the two, Rizzo also claimed that the ISAF mission was 'a reflection' of NATO transformation. The nuance here is unmistakeable: according to Rizzo, NATO's transformation was the key driver of the mission, and increased involvement in Afghanistan was a consequence of the demands of transformation and multi-national cohesion. One may infer from his statement and ordering of rationales therein that the needs of the Alliance shaped the mission; in fact, to a degree the expansion of the Afghan mission was conceived in terms of testing the mettle of the Alliance. Simple chronology suggests this to be the case, since NATO's transformation agenda was set in motion just months prior to its decision to take over ISAF. Evidence of this ordering of priorities appeared again in a speech given by former NATO Secretary General Lord George Robertson in October 2003 when he listed the reasons for NATO's new role:

> if we abandon the people of Afghanistan, the international credibility of every NATO nation, and of the Alliance itself, will lie in tatters. The risk from mass terrorism will increase exponentially. And the prospect of ending the scourge of Afghanistan sourced drugs on the streets of Brussels and other European cities will vanish. (Robertson 2003, online)

Again, the self-preservation of NATO appears as the foremost rationale for expanding the ISAF mission, followed by the mitigation of international terrorism and eradicating the narcotics trade. Interestingly, the well-being of the Afghan people was not listed as a reason for involvement by Robertson, just as international terrorism and the 'scourge' of heroin were not mentioned by Rizzo two months earlier. The theme of NATO's credibility being presented with a fundamental test in the form of Afghanistan remained constant for over a year, with Robertson's successor Jaap de Hoop Scheffer employing this argument as late as October

2004 (Scheffer 2004, online). What this suggests is that institutional demands – specifically the maintenance of Alliance cohesion in the face of a major military operation, and therein the validation of NATO as a post-Cold War institution – were of paramount importance, or at the very least were equal in importance to the task of stabilising Afghanistan. It was not a pure form of 'strategic logic' – the threat posed to Europe by Afghanistan – but multi- or transnationally informed 'institutional dynamics', specifically 'Europe's political commitments and interdependencies', as re-configured in light of the Iraq affair, that defined the NATO decision and shaped the course of ISAF operations (King 2011: 25). Indeed, it would not be until two years after the NATO takeover that a senior British statesman would explicitly relate Afghan instability to British 'national' security. In a speech to the Labour Party Conference in Brighton following the July 2005 bombings in London, then-Chancellor Gordon Brown noted how the denizens of the capital had moved in 24 hours from the 'joy' of winning the prize of the host city for the 2012 Summer Olympics to

> the horror of homegrown suicide bombers maiming and killing fellow British citizens … let no one doubt that we will spend what it takes, bear each and every hardship, endure each and every sacrifice … both internationally in Afghanistan and Iraq and at home, we will at all times have the strength and resolution so that there is no hiding place for terrorists – or those who finance terrorism – and so we will protect and defend the security of the people of this country. (Brown 2005, online)

Brown's statement is interesting on a number of levels. Firstly, and most obviously for our purposes, it signalled the emergence of a nascent 'national security' element to the counter-terrorist narrative as a consequence of the first successful Islamist terrorist attack in the United Kingdom. The bombings exposed the vulnerabilities of the national security apparatus in a manner that left few in any doubt of the dangers posed by religiously-inspired terrorism. In making this rhetorical move, however, Brown did not jettison the overarching logic of the internationalist vision expounded by Blair. For the sake of narrative coherence it was fortunate that Blair's approach was flexible, because the facts of '7/7' did not directly support it. The perpetrators of the bombings were British citizens, not Afghans or Iraqis; they had visited not Iraq or Afghanistan in preparation for their activities but Pakistan – an ally in the War on Terror; and it was likely that Britain's involvement in Afghanistan and Iraq had actually increased radicalisation of Muslim communities in the UK and, therefore, the likelihood of terrorist attacks on British soil. These issues did not prevent Brown from linking the attacks to al-Qaeda activities in Britain's theatres of operation and, as a consequence, from equating Britain's collective security commitments in those countries as being directly in the country's national security interest. The mere fact that the London bombers were 'inspired' by al-Qaeda was sufficient to make the claim that they were connected to the terrorist 'network' and, therefore, indicative of an existential struggle that knew no geographical or administrative limits but, on the contrary, was defined by norms. This fit neatly into Blair's dialogical view of a battle of 'values', which allowed judgements on the strategic worth of occupying or invading states on grounds of collective security to be unhindered by the fact that the only terrorist attacks successfully carried out in Britain since 9/11 were perpetrated by residents of Britain. Tellingly, however, Brown's national security approach was not repeated by Blair when he took the stage at the Party Conference the following day. Instead, Blair remained true to his internationalist roots in reminding the audience of the rationale for Britain's leading role in the War on Terror:

> I never doubted after September 11 that our place was alongside America and I don't doubt it now. And for a very simple reason. Terrorism struck most dramatically in New York but it was aimed then, and is aimed now, at us all, at our way of life. (Blair 2005, online)

For Blair, justifying collective security principles required the maintenance of a discourse that eschewed 'national' considerations and placed the conflict in ideological and normative terms that emphasised the essential unity of alliance members as an international community of like-minded democratic actors. This is not to argue that one could not independently read a 'national' aspect into Blair's statements and speeches; on the contrary, the strength of Blair's position was that it could bridge the divide between international obligations and domestic political pressures with limited contradiction. Rather, the issue was that Blair's internationalist outlook predicated that the first of these (foreign relations) could not be sacrificed for the second (domestic political expediency). As such, where pressures emerged from within the UK that challenged Britain's liberal internationalism, Blair's response was limited in that the normative narrative edifice built around that principle was inflexible: alliance commitments were for his Government a 'red line' that could not be sacrificed by reversion to a populist realist revisionism centred on a traditional view of 'national' security. In retrospect, Blair's Brighton speech appears as the high point of his particular brand of internationalist rhetoric. His achievements up to then were significant: an unprecedented third consecutive victory for Labour in the 2005 General Election, leadership on the world stage unsurpassed by his European counterparts, and contribution to the undertaking of an historic parliamentary election in Afghanistan. Britain's entry into Helmand in 2006, however, would mark the beginnings of a shift change in emphasis of the counter-terrorist policy narrative away from international security and towards a more restricted form of national security.

Phase III: 2006–2009 – From Afghan Democracy to British Security

Preparations for Britain's participation in ISAF's move to the south of Afghanistan in early 2006 were accompanied by a rhetorical dimension that initially reaffirmed the Blairite conceptualisation of foreign affairs as one defined by interdependence, solidarity among NATO allies, and a strong vein of Kantian ethics. Much of the spokesperson duties to this end were performed by Defence Secretary John Reid, a close ally of Blair's, who spoke publicly on the subject of deployment at least 11 times between June 2005 and the beginning of the operation in April 2006. Reid explained the role of Task Force Helmand as a reconstruction and security-building mission, and in doing so emphasised the distinction between ISAF operations and those of the American-led Operation Enduring Freedom, which were delimited to counter-terrorism operations (Hansard 2005d). He took care to frame Britain's operations within the multilateral framework established by Blair, declaring in January 2006 that the drive into the southern quadrant of Afghanistan was 'truly… an international, multinational effort' (Hansard 2006a). In a joint press conference with Donald Rumsfeld in February, Reid articulated the multilateral approach in a manner reminiscent of Blair's Chicago speech, stating that

> [t]here's a reason that NATO is involved in Afghanistan and it's that this is the 21st Century. Problems are not specifically nation state problems, they're not specifically even regional problems and in many instances they're global problems. It requires an alliance like NATO to recognize and adjust and face the challenges that exist in this new century. (Reid 2006, online)

In Reid's view, then, Helmand represented a multilateral, even transnational response to the problem of failed states, of which the British were one component within a transnationalised defence edifice working for the end of global security. The NATO alliance's full participation in the stabilisation and reconstruction of Afghanistan was necessary because all were subject to the threat posed by instability in that country to their own security, and because all nations' security were at risk from terrorism on the basis of their shared values, the security threat was a global one rather than a national one. Because the threat of terrorism was considered indivisible between NATO states, it followed that counter-terrorism, whether pursued directly in the form of capture-or-kill missions, or indirectly through stabilisation and reconstruction activities, was the responsibility of all NATO states. In this sense, Reid was outlining a classic interpretation of collective security logic. Ten days after his press conference, in an interview with the BBC, Reid linked the national security of the United Kingdom to that of the success of the collective security mission by connecting the indivisibility of national security to the solidarity necessary for the success of the mission.

> We)re there because Afghanistan under the Taleban and Al Qaeda, was the place where not only were the plans to murder thousands of people in New York prepared, but they were actually trained for, and launched. And if Afghanistan is ever to slip back into the hands of the terrorists again then we would be just as much in that front line of threat as the people of, the thousands who tragically died in New York. (BBC 2006b, online)

Reid did not entirely comport with Blair's totalising counter-terrorism approach, however. In the same BBC interview he drew a subtle distinction between Afghanistan and Iraq, stating that Britain 'went in there unlike Iraq with the united world community through the United Nations' (BBC 2006b, online). This was presumably an attempt to enhance the legitimacy of the Afghan mission *vis-à-vis* Iraq in the eyes of a sceptical British public by presenting its multilateral, UN-sanctified credentials, and possibly also represents a nod by the Defence Secretary to the British military's dissatisfaction with Iraq and their desire to shift the locus of their operations out of Basra and into Helmand. This move represented a sharp diversion from the Blair approach of treating Iraq and Afghanistan as two fronts, of equal legitimacy, in the same war, and can also be seen retrospectively as the beginnings of a narrative line that prioritised the comparative treatment by British officials of Afghanistan, contra Iraq, as 'the good war' (Kavanagh 2012: 50).

By the autumn of 2006, the framing of the counter-terrorist policy narrative had begun a period of gradual change, most notably in the altering of the referent object of security from the 'global' to the 'national'. This change of emphasis was likely the result of rhetoric meeting reality on the ground in Helmand: for all the pre-deployment talk of reconstruction and development, the environment encountered by British forces from April 2006 was primarily combat-based, involving a series of violent clashes between insurgents and thinly-spread British troops and culminating in a negotiated cease-fire and tentative truce in Musa Qala (King 2010b: 317–18). From the start of operations to the end of August, 16 British troops had been killed in the fighting, compared with just four in the entirety of the campaign between 2001 and 2005 (three of which were non-battle related). Naturally, this discrepancy between rhetoric and reality was latched upon by an increasingly critical British press, which published articles questioning the coherence and logic of the mission aims of reconstruction and counter-narcotics in such an environment (see, for example, Jenkins 2006). Recourse to a narrative that linked British military activity and casualties to the protection of British citizens at home can be seen in this light, as with the 45-minute claim, as a rather instinctive and defensive narrative position to take in the face of mounting criticism.

The beginnings of this trend were tempered in their attempts to balance the high-concept normative basis for a global or collective security imperative with the rhetorical impulse for national security. Blair spoke at the September 2006 Labour Party Conference in Manchester of the dangers of withdrawing prematurely from Iraq and Afghanistan as akin to 'a craven act of surrender that will put our future security in the deepest peril', but remained resolute in his internationalist vision by noting that although '[t]he British people today are reluctant global citizens', it was up to his Government and the Labour Party to 'make them confident ones' (Blair 2006, online). The Opposition leader, David Cameron, spoke in similar terms the following month, claiming that leaving Afghanistan will 'create another cocktail, another hotbed of terrorism. That is bad not just for Afghanistan but bad for the world. Bad for us here … in the United Kingdom' (Cameron 2006, online). Balancing these demands was not necessarily problematic since they were based on a hypothetical claim that had long been a staple of the British (and international) counter-terrorist narrative, indeed one that underpinned the entire rationale of ISAF stabilisation activities: that an unstable Afghanistan would, as a matter of course, return to the state of affairs pre-9/11. Of course, this assessment was unfalsifiable, but had the trauma of 9/11 to support it. It was also uncontroversial in its ubiquity throughout the policy positions and public statements of members of the NATO alliance. An early exception to this rule is found in an interview conducted by the BBC with Shadow Defence Minister Liam Fox, who prioritised national security as the primary rationale for involvement in Afghanistan for the first time:

> [i]t's absolutely essential that for the future functioning of the UN and its authority, but more importantly for the cohesion and the reputation of NATO, that we have success in Afghanistan. And it has a direct impact on this country's national security, *which is top of the list* [my emphasis]. (Fox 2007, online)

Fox paid homage to the core interest of the United Kingdom in maintaining the NATO alliance but also categorically stated the pre-eminence of British interests by introducing an explicitly 'national' security argument. At this stage, however, Fox's statement represented something of an exception to the rule. So long as Blair remained Prime Minister, the rhetoric of both his ministers and his political opponents generally maintained their consistency, speaking to the normative basis of the mission and to generic security concerns, but almost always within a wider context of global or collective security. What is remarkable about Fox's interview was that his ostensible purpose of appearing on the programme was to criticise what he perceived to be the lack of adequate 'burden-sharing' within NATO, noting that the British, along with the Americans, Canadians, and Dutch were performing a disproportionate amount of the most dangerous work in the country, while NATO members such as Germany, who operated under strict national caveats dictating their rules of engagement, remained in relatively benign northern parts of Afghanistan. The perception that British personnel were incurring higher casualty rates than many of their continental counterparts could not have helped the narrative device of multilateralism, particularly within the Conservative Party which, whilst populated by a number of internationalist-minded parliamentarians, did not share Blair's vision for 'relocating the national interest' within the international community. Citing 'national security' therefore made ideological sense and had the fortuitous added benefit of political expediency.

Gordon Brown's accession to the role of Prime Minister in May 2007 was accompanied by an enhancing of the national security discourse within the counter-terrorism narrative. In general terms, Brown sought to distance himself and his policies from that of his predecessor, moving away from the grandiose rhetoric of Blair and toning down the 'special relationship' with President Bush (Dunn 2008). This appears to have been a very conscious move: it may be argued that Brown's motivation as framing himself as a contrast to Blair and with the goal in mind of mitigating the damage done to Labour by Iraq. As far as concerns

Brown, however, Iraq was something of a moot point by the time he took over from Blair, as the wheels were put in motion by the military chiefs in the early 2007 to begin drawing down from Iraq and reallocating troops and resources to Afghanistan (Seldon and Lodge 2011: 26). Nonetheless, as David Dunn (2008: 1135) noted, the trend of political discourse under the Brown ministry was to disassociate the Government from the negative implications of the Blair era, specifically the popular perceptions of 'poodleism' and the acquiescence of the British state to the demands of the Bush Administration on Iraq. This was arguably an element of the counter-terrorism policy narrative for Afghanistan that had become toxic under Blair by its association with Iraq, and therefore required British policymakers to move away from emphasising the collective security dimension of the Afghan mission. By asserting Britain's independence from the United States, it was therefore perhaps a natural corollary that the language of national security would become more prominent. A result of this shift in tack was that members of the Cabinet, freed from Blair's influence, were quick to point out the differences between Iraq and Afghanistan in public statements, focusing on the importance of the latter by implicitly casting doubt on the security relevance of the former. For example, in an interview in July 2007 laying out the Brown Government's foreign policy agenda, just two months after Blair's departure, Foreign Secretary David Miliband made the following remark:

> I hope we can move to a situation where we don't have Iraq and Afghanistan in the same breath... they are different conflicts with different dynamics. I think the situation in Afghanistan is a central aspect of national security. If you think about 9/11. (Miliband 2007, online)

Moving to separate Afghanistan from Iraq by emphasising the national security element set in motion a chain of events throughout the remainder of this phase which would culminate in the pre-eminence of the national security argument above all others. Throughout the remainder of 2007 and the first half of 2008, the *modus operandi* of official statements on Afghanistan was to highlight the multi-faceted 'comprehensive approach' to securing a democratic Afghan state and society. National security was an important, but still not paramount, element in the counter-terrorism narrative. Miliband in particular embarked on a speaking tour promoting, in a manner reminiscent of Blair, the British effort in Afghanistan as one that 'symbolises our dual goal of protecting our national security and promoting human rights' (Miliband 2008, online). In October 2008, however, with the departure of Des Browne and arrival of John Hutton as Defence Secretary, the narrative rapidly altered to one that placed national security, and specifically 'national security interests', at the forefront. According to Seldon and Lodge (2011: 207) Hutton's appointment as Defence Secretary in early October 2008 was met with some consternation by the Foreign Office and the Department for International Development. During Browne's tenure, Seldon and Lodge argue, a 'consensus had existed' between him, Douglas Alexander and

Miliband of 'finding a political solution and building up Afghan institutions... rather than laying too much weight on a military defeat of the Taliban'. Hutton, they contend, was perceived by Brown as 'the voice of the military in Cabinet, rather than the voice of government in the MoD'. In this reading, Hutton could hardly have been more different from Browne. The narrative change appears to have been near instantaneous: whilst Miliband and Alexander had for the previous year spoken frequently of the need for a comprehensive approach, Hutton wasted little time in making several statements and speeches arguing for more robust military activity and resourcing.

In November, Hutton gave a speech outlining his view of Afghanistan as representing a 'vital national security interest'. He began by addressing the reasons for going into Afghanistan, arguing that the September 11, 2001 attacks in the United States were an 'attack on our national interests', and claimed it was 'a crime by Al Qaida against the entire civilised world' (Hutton 2008, online). His remarks can be interpreted as being somewhat paradoxical and, as such, quintessentially in keeping with liberal normativity, insofar as he attempted to synthesise a realist appraisal of state interest with a liberal internationalist assertion of interdependence. Whilst couching 9/11 and the war in Afghanistan in a distinctly 'national' context, Hutton essentially echoed the same logic as Blair's internationalist rhetoric seven years previous, presumably because he recognised that one cannot say a national interest has been attacked as a result of an attack on a foreign city without making reference to institutional and/or liberal norms implicit in collective security arrangements. Yet, in doing so, the implication is that one naturally mitigates much of the 'national' aspect of the interest itself. Given the distinctiveness of such a turn of phrase, it is worth querying why Hutton might have taken this narrative approach and, more importantly, why Hutton's national security narrative became entrenched within the Brown Government thereafter.

There are several possible reasons we can consider. The first relates to collective security demands: specifically, the recognition by the British political establishment of changing circumstances in the United States and the fact that both presidential candidates in the 2008 election were 'pro-surge' (Defence Committee 2008: online). Under such a surge, Britain would be expected to contribute even more resources to the Afghan campaign, which concomitantly would increase the likelihood of more fighting, more casualties, and more negative media attention. As such, reconfiguring the conflict as nationally-focused and about British security and interests would ameliorate the potential for government officials to contradict one another. A second issue relates to the challenges of producing a convincing stabilisation narrative that aligned with the broader expectations of civil society. A narrative that focused on developmental and political issues whilst accounting for the costs associated with deteriorating security on the ground was no simple task, particularly given the media's tendency to be only interested in reporting on security. The view in Whitehall was that British messaging had become too narrow in scope, and that an enhanced national security narrative would be useful in 'lifting it back out of Helmand' – in other words, by emphasising less the specifics of development and the hardships therein, and more the broader theme of nation-wide stabilisation as the means to the end of counter-terrorism (Personal Interview 2013). This view has been corroborated in an interview with another source, who suggested that the narrative had been simplified for the purposes of consistency of message to the British public, because the narrative that pre-existed that of counter-terrorism was overly complex and tended to confuse domestic audiences (Personal Interview 2013).

A third potential reason is party political: the Brown Government's adoption of the national security discourse from the Conservative Opposition (particularly from Cameron and Fox), who had been arguing with some consistency for two years – and with greater intensity in the summer and early autumn of 2008 – for a national security-based narrative for the conflict. This in turn may have been the result of the close links between Hutton and the military, and between the Conservatives and the military. A fourth contributing factor is economic, that is, the impact of the global financial crisis on public spending, and a concordant recognition by Brown that the necessities of spending cuts would potentially impact the scope and breadth of what could be achieved in Afghanistan, thereby making it prudent to focus the narrative on a core issue of national security (Cornish and Dorman 2009: 248). Whatever the reasoning (or combination thereof) behind the narrative transition, the language used in Hutton's speech was new, particularly the term 'national security interest':

> the decision to stay was based on a hard-headed assessment of our clear national security interest in preventing the re-emergence of Taleban rule or Afghanistan's decline into a failing state again. Either of those outcomes would have allowed Al Qaida to return and recreate their terrorist infrastructure. (Hutton 2008)

Hutton's use of the term 'national security interest' was, in fact, a first for the narrative for the war in Afghanistan. There is little that is 'clear' about it upon first inspection; rather, it seems to be merely a hybridisation of the terms 'national security' and 'national interest'. When one talks of national *interests* it is important to recognise that the referent of those interests is, at the fundamental philosophical level, disputed; as Arnold Wolfers noted, 'when political formulas such as "national interest" or "national security" gain popularity they need to be scrutinized with particular care. They may not mean the same thing to different people. They may not have any precise meaning at all"' (Williams 2012: 8). While we do not know precisely what Hutton meant by it, or whether he meant anything by it at all – it may well have been an uncontemplated rhetorical idiosyncrasy – the term itself stuck, and would be frequently employed by the Brown and Cameron ministries in the years that followed (Hansard 2011; Defence Committee 2013:Ev71). A reasoned inference may be that the use of the term 'national security interest' was a subtle attempt to reconcile the divergent political philosophies of realism and liberal internationalism. To say that a stable Afghanistan is 'in the national interest' does not necessarily imply that it is a security threat to the United Kingdom; it may be so, but may also be in the national interest because of collective security demands or so that Britain's place in the world remains at its current disproportionate status. Equally, to imply Afghanistan is a national security issue outright (as opposed to a 'security interest') would be to give rise to an understanding that it poses a 'direct' threat to the UK, thereby taking on a rigidity that contrasts with a national security interest. Indeed, examinations of the rationale for participation in Afghanistan of NATO member states reveals that national security is often subordinate to alliance cohesion concerns since, rather than enhancing national security, many NATO members were of the opinion that involvement made them less secure (*The Guardian* 2009).

Specifically, to say that Afghanistan represents a national security threat to the United Kingdom requires some evidential basis that can be easily deployed to substantiate the argument. A 'national security interest', however, does not, since its hybridisation of interest and security indicates a flexibility of application. Precise causality has seldom been forthcoming by state officials in this regard, although this may simply be a reflection of the clandestine demands of British intelligence institutions. Whatever the truth may be, the lack of exact causality in Afghan messaging was partially remedied by Hutton, however, by resort to personal conviction and hypothetical supposition:

> We undertook military action in Afghanistan because this was the base from which Al Qaida leaders, through the sanctuary offered by the Taleban, were planning and directing major terrorist operations throughout the world – operations that would, without any doubt at all in my mind, have been aimed at the UK…let us be clear: if they had the means no moral compunction would restrain the unleashing of those kinds of destructive forces on the streets of Britain. (Hutton 2008)

Hutton's line of argumentation could point to the terrorist attack of 7 July 2005 (as well as the foiled attempt of a fortnight later) to vindicate his view with some plausibility. However, phrases such as 'without any doubt at all in my mind' and 'if they had the means' represent in the main unfalsifiable personal opinions; whether these opinions were supportable by evidence is not clear, but any such evidence was not provided in Hutton's speech. As with the term 'national security interest', such hypotheticals would become commonplace in the latter stages of the campaign. Despite the longevity of this narrative device, the limitations of such an approach were exposed relatively early – in October 2008 – during a Defence Committee hearing, in an exchange between Committee member and Conservative MP David Heathcoat-Amory and Miliband:

> Heathcoat-Amory: Frankly, recent international terrorist threats, as far as we can assess them, have not been solely or even mainly derived from Afghanistan.
>
> Miliband: That is because we are there.

Heathcoat-Amory: It is desirable to have a democratic, strong Government in Afghanistan, but that is too far removed from our own security to be a realistic or precise war aim that we can judge that we have met and so end the campaign.

Miliband: Obviously, that is a ridiculous argument. You are saying that there is no reason for us to be there. I am saying that we are there to prevent the Taliban from overthrowing the Government and thereby providing a home for al-Qaeda. You are saying that there is no evidence that al-Qaeda is a threat in Afghanistan. Precisely, but that is because we are there. (Defence Committee 2008)

Heathcoat-Amory's questioning shed light on the paradoxical nature of the suppositional position taken by the Brown ministry, namely that Afghanistan is a national security threat because of 9/11 and, because of this past reality, if Afghanistan were to fall back into the hands of the Taliban there would be a substantively higher risk that al-Qaeda would return and seek to attack Britain. However, because Britain and its ISAF partners were in Afghanistan this threat was being contained, and as such Afghanistan did not currently represent an immanent security threat to the United Kingdom. Heathcoat-Amory's criticism is indicative of exactly the kind of rhetorical trap a Government messaging policy of positing Afghanistan as centrally a counter-terrorism mission can fall into. Similarly, such a position opened up to questioning the validity of pursuing a democratisation agenda in Afghanistan if the sole aim of Britain's involvement in the conflict was national security. Because the focus had shifted from international to national security, scrutiny increased on the democratic and development aspects. A consequence of the discourse of national security was the emergence of a situation where stabilisation was relegated to the status of a means to the end of a nationally-focused counter-terrorist agenda. This would set the trend for the final five years of Britain's participation in the Afghan campaign and provide the catalyst for the undoing of the stabilisation narrative.

Phase IV: 2009–2011 – Consolidating National Security Rhetoric

During the fourth phase of the Afghan conflict the frequency of the national security line of inquiry increased dramatically. Substantial revisions and expansions of the UK's counter-terrorist narrative took place during this period, including the incorporating of Pakistan into the foreign internal defence stratagem, reflecting President Obama's March 2009 reorientation of US policy to the 'Af-Pak' paradigm (The White House 2009a, online). Obama's announcement of Af-Pak defined the mission as the elimination of al-Qaeda safe havens along the border of the two countries, thereby validating the British political establishment's national security discourse set into motion the previous year. It also provided a greater delimitation of the mission's parameters and confirmed a shift in US policy and resources away from Iraq, undoubtedly improving relations with America's increasingly war weary continental European counterparts. Indeed, by late 2009, with the announcement of Obama's Afghan 'Surge', practically all major NATO members had announced an increased troop contribution to the conflict. The way the governments of these member states spoke about the importance of the mission differed dramatically from that of Britain, however. Sarah Kreps (2010: 203–9) notes that practically all of Britain's major NATO allies (excluding the United States) – Germany, France, Canada and Italy – employed a rationale that affirmed, more than any other factor, the importance of multilateralism and alliance cooperation, not national security concerns, as the prevailing reason for sustaining a troop presence in the country.

Following the Hutton speech and the introduction of SC processes in the Ministry of Defence, the national security rhetoric coming from Government was emboldened. Labour ministers jettisoned language that emphasised indirectness of threat, and frequently employed terms that evoked, for the first time, a direct link between instability in Afghanistan and the potential for 'terror on our streets' in the United Kingdom. Between April and November 2009, for example, Brown spoke of the potential for Afghan terror plots to be carried out, via a 'chain of terror that links the mountains of Afghanistan and Pakistan to the streets of Britain' (Hansard 2009a). This rhetoric hit new heights over the summer of 2009, with Brown, Miliband, newly appointed Defence Secretary Bob Ainsworth and Armed Forces Minister Bill Rammell taking the lead in arguing the national security case for sustained involvement amidst some of the fiercest fighting (and fiercest media criticism), and the highest summer troop death toll (57 between May and August) of the conflict to date. The Af-Pak theatre was given various metaphors to drive home this new enunciation of threat, including the 'crucible', 'incubator', 'epicentre' and 'nexus' of terrorism (Hansard 2009a, 2009b). To support this claim, ministers repeatedly referred to the fact that 'three-quarters of the most serious terrorist plots against the UK have germinated in the mountains between Afghanistan and Pakistan', but also – counterproductively for their purposes – that 'seventy percent of the terrorist attacks planned in the UK have links back to Pakistan' (Miliband 2009a, online). This clearly implied that the vast majority of terrorist plots were concocted in Pakistan, whereas relatively few (something to the order of five per cent) were conceived in Afghanistan. Public officials such as General David Richards would address the concerns of those who doubted the Government's national security reasoning on the basis that Pakistan was more of a threat to British national security than Afghanistan by practicing a rhetorical *argumentum ad ignorantiam,* ostensibly to shift the burden of proof away from the counter-terrorism narrative and onto its detractors by securitising the debate:

> There are some that question whether a victory for the Taliban would lead automatically to a resurgence of al-Qaeda inspired terrorism in the UK and other western countries. I ask you this, can you be certain that it would not?. (Richards 2009, online)

Richards' statement can been seen in one sense as eminently sensible in its neat articulation of the risk paradigm: can the intellectual and strategic poverty – and political necessity – of positing unfalsifiable scenarios be overcome by simply rejecting it outright on the basis that it is a logical fallacy? Doing so would no doubt in his view represent little more than an abrogation of a state's first duty to defend its citizenry against attack. Indeed, there seems little justification, given the context of British operations in Afghanistan in 2009, in blaming public officials for feeling duty-bound to highlight national security demands.

Rather, it would seem that the logic of emphasising national security was unassailable given the situation with mounting troop casualties and dwindling public support for the mission, which by 2009, according to opinion polls, amounted to more than half the electorate 'questioning their country's mission' in Afghanistan (Angus Reid/BBC 2009, online). By focusing on the national security element of operations, the Government sought to provide the strongest and most culturally embedded justification for participation they had at their disposal. The problem with this narrative approach was that it quickly confounded itself as a result of political strategic developments taking place in the United States. Specifically, it is worth considering the timing of the Government's shift to the counter-terrorism narrative from more Afghan-centric arguments and the emergence of 'national security' as the focal point of all policy narratives, which coincided in late 2008 with the strong showing of Barack Obama in the run-up to the US presidential election. Obama's official position on the War on Terror was to move away from the dogmatic thought processes that led America and its allies into Iraq, and focus instead on the core aspects of US national security. One may therefore postulate that the British Government's narrative shift toward national security was in part in anticipation of Obama's likely electoral victory. If this was a contributory factor, it would be quickly complicated by Obama's strategic reset in mid-2009 to incorporate Pakistan into counter-terrorism policy. Britain's inclination to follow America's lead in counter-terrorism and emphasise the 'national security' element was effectively undermined by doing so, as the narrative work done to establish Afghanistan as the primary security threat to the United Kingdom was undone. Indeed, the result was that the Government's Foreign Office Minister, Lord Malloch-Brown, would claim that it was now Pakistan, and not Afghanistan, that represented the primary national security threat to the United Kingdom (Telegraph 2011, online). This inconsistency indicated a larger problem, as well. By placing national security over and above regional or global security – and therefore over the stabilisation requirements for Afghanistan – the Brown Government inadvertently undermined the logic of the ISAF mission altogether, since it could no longer appeal to the non-counter-terrorism-related accomplishments. By placing counter-terrorism at the core, it therefore called into question the relevance of Afghanistan in comparison with Pakistan and elsewhere. As Brown's aide Matt Cavanagh would later opine, if the mission was solely about British national security, could the UK not 'achieve that in a different way with fewer troops and casualties, and less money? Indeed, if it's all about al Qaeda, why are we in Afghanistan at all, rather than Pakistan, or even Somalia and Yemen?' (Cavanagh 2010, online).

The contradictions implicit in reframing Afghanistan as primarily about UK national security were evidently elusive to Brown's ministers, however. In their efforts to make the shift in narrative toward a narrow, nation-centric conception of counter-terrorism, they created ramifications for nation-building to the point where, in order to make the point that Afghanistan was at root a mission of national security, they seemed to disparage the notion that there were any non-security aims to the mission whatsoever:

> the point I think we consistently need to make clear is that we're in Afghanistan, not because we want to make the world a better place in that part of the world, but because fundamentally this is about our national security and our national interest. (Rammell 2009b, online)

This statement rather plainly contradicts Blair's rhetoric outlined in the first phase of the conflict, to say nothing of the aims of the stabilisation mission under Brown. The counter-terrorist narrative as originally conceived by Blair had the transformation of Afghanistan into 'a better place' at its very core, and explicitly eschewed any idea of a rationalist prime directive of security imperatives by placing the national security interests of the United Kingdom as an indirect goal at best. Perhaps this was merely a reflection of the perception that security had not improved as much or as quickly as Blair anticipated: by late 2009, when Rammell made the comment just quoted, the security situation in Afghanistan remained perilous, with most of the developmental benchmarks as set out in the Afghan Compact of 2006 unachieved and a presidential election left in disrepute by allegations of widespread fraud (Miller 2011: 56; Asia Foundation 2011: 31). Indeed, Rammell was not alone in his understanding of the purpose of Britain's mission: the frequency of similar statements suggests this downplaying of non-security issues was an agreed upon line of Government communication policy. The extent of this reframing was further exposed by Miliband's November 2009 speech on NATO's future direction in Afghanistan, in which he demarcated the policy aim of preventing the country 'being used again, by Al Qaida under the umbrella of Taliban rule, as a launching pad for international terrorism' from 'the strategic plan – to support the development of Afghan institutions to deliver this goal' (Miliband 2009c, online). This was quite a different position from the Miliband of 2007 and early 2008, who spoke of 'protecting national security and promoting human rights' as being a 'dual goal' and of the betterment of Afghan society as essential to British interests. On the other hand, the Miliband of 2009 demonstrated significant conceptual progress in his ability to demarcate policy from strategy. Interestingly, both Rammell and Miliband's comments came just weeks after Shadow Defence Secretary Liam Fox appeared on the BBC to lambast the Brown Government's communication efforts and their tendency, in his view, to muddle messages of security with those of Afghan development:

> you have to be very clear about what the military mission is and what the subsequent mission about reconstruction and so on, which is for the international community to do. Because… the public understand sending our troops to fight for our national security. They don't understand the concept of fighting for an education policy on the other side of the world. (Fox 2009, online)

Fox's comments appear to once again have had a significant impact on, or at least anticipatory of, Labour messaging efforts. They also represent a concise summarisation of British SC doctrine as found in *Joint Doctrine Note 1/12*, which stated that '[t]he logic of the strategy and its appeal should be compelling and easily understood … in its totality should be simple, or at least capable of explanation in simple terms' (DCDC 2012: 1–4). This involved focusing on public opinion by relating the message to the public in terms with which they were familiar. Fox clearly believed that this required framing the conflict with reference to the United Kingdom's national security demands. The stated aim of strategic communication in *JDN 1/12*, to 'advance national interests', was a theme repeated by Chief of the Defence Staff Jock Stirrup in December 2009 when he remarked that '[a]ll we do at the tactical and operational level needs to be rooted in good strategic soil, and therefore in our national interest' (Stirrup 2009, online). In order to root policy in the national interest, however, the Government were evidently compelled to renege on the internationalist vision for Afghanistan and the core element of stabilisation which, while still rooted in the national interest by degrees of separation – specifically, transnationalised defence policy – was not as effective in rhetorical terms to the concept of 'national security' as the revised, nation-centric counter-terrorism narrative.

The difficulties of maintaining a coherent narrative line on the purpose of Britain's activities in Afghanistan ultimately resulted from a transnational defence policy, on the one hand, and a national defence rhetoric, on the other, that were clearly at odds. In 2011, *The Daily Telegraph* published a Government communique from mid-2009 made available for public scrutiny by the whistleblowing website *Wikileaks* that clearly showed Britain's strategic decision-making was dependent upon, according to the communique's authors, the 'outcome of the U.S. assessment of operations in Afghanistan as a prerequisite for determining key aspects of British strategy in Afghanistan' (*Telegraph* 2011, online). This way of thinking demonstrates in no uncertain terms the political realities facing Britain's ability to devise strategy; that is, it cannot be devised in any substantive way independently of strategies pursued by its primary partner in collective security operations. A logical consequence of this dependency, of course, is that the narratives explaining the strategies were also contingent upon the prevarications of American foreign policy decision-making. Given that this nuance must not have been lost on Stirrup, one may infer that rooting what Britain does in 'good strategic soil' and in 'the national interest' means either that Stirrup considered such soil to be of reasonable quality despite it being located without the United Kingdom, or that – as the opening chapters of this book argued – that for senior policymakers and military officials the national interest is either understood as largely inseparable from the interests of the collective security framework or, more worryingly, is uncritically assumed to exist independently of transnational socialising processes. Whichever of these options best applies, they all speak to the transnational dilemma at the core of British policymaking and strategic thinking: that Britain is inexorably at the mercy of the will of the collective security system (specifically that of the United States) in order to realise its interests.

Curiously, despite 18 months of pressing the 'national' security element in Britain's collective security operations in Afghanistan, some in the Government spent the spring of 2010 attempting to restore the internationalist dimension of the counter-terrorist narrative in a manner reflective of Blair's original genre guess. In April 2010, Ainsworth released a 'Labour vision for defence policy', presumably designed to differentiate Brown Government from the Conservatives in the run up to the 2010 General Election. In his speech, Ainsworth highlighted many of the themes that Blair had promoted and that had been cast away from the narrative in the years following his departure from office:

> today we are fighting in Afghanistan as part of an international coalition to protect global security from the threat of terrorism. Labour believes it should always be the role of powerful nations, such as our own, to support a rules-based international system. We believe in universal human rights, democracy and multilateralism... We are in Afghanistan to ensure that the country cannot again be used as a base to export terrorism that is a proven threat to our citizens. (Ainsworth 2010, online)

Ainsworth's attempt to encapsulate the core elements of Labour's defence policy reads as a *post-hoc* corrective to the steady move away from internationalism under Gordon Brown. In most of the significant alterations of the counter-terrorism narrative towards a more avowedly 'national' focus on defence, the Conservatives were chronologically prior in making the case. As such, it is perhaps unsurprising that, following the electoral defeat of the Labour party and the coming of the Conservative-led Coalition Government in May 2010, the national security-based counter-terrorism narrative continued in much the same vein as it had under Brown. Equally unsurprisingly, they also encountered the same problems as the Labour Government in making the case for Afghanistan as a counter-terrorism mission, with ministers being frequently questioned in the media about the veracity of such a narrative approach given that the majority of terrorist plots against the United Kingdom were planned or carried out by British nationals (Rammell 2009a, online; Fox 2011a, online). A narrow focus on national security also drew criticism from supporters of the stabilisation agenda – particularly those interested in the advancement of women and girls in Afghan society – in response to Fox's comments of 'not fighting for an education policy' (BBC 2010, online; Fox 2010, online). The logic of staying in Afghanistan until it was stable was equally criticised, however. A month after the election, Ainsworth appeared on the

BBC to defend the national security logic of remaining in Afghanistan and was subjected to a scathing critique by *Guardian* columnist Zoe Williams:

> [t]here's never going be a position where we can say there's absolute national security, still less is there ever going be a position where you can say 'we in the UK are nationally secure against terrorist action'. It's completely ridiculous... it's unprovable, it's unfalsifiable, it's absurd, so really what you're talking about is government versus public opinion, and that is, that is just a matter of time really. (BBC 2010, online)

Whilst uncharitable, Williams' point of view encapsulated all of the weaknesses of the national security-based counter-terrorism narrative, and in doing so put the narrative in juxtaposition with the strategic communication imperative of having a 'compelling' and 'easily understood' logic. Indeed, the weeks following this exchange would prove pivotal in addressing the content of Williams' argument. Just three days later, Cameron would address the Commons on Afghanistan for the first time, reaffirming the national security rationale of Britain's continued involvement in the conflict and warning against premature drawdown by arguing that any assessments for such action 'must be done on the basis of the facts on the ground, not a pre-announced timetable' (Hansard 2010c). At the G8 Summit in Toronto at the end of June, however, the call for strategic patience had been reversed, with Cameron announcing – unilaterally and without consultation with Cabinet, according to one source – a timetable for withdrawal by the end of 2015 (Personal Interview 2013; Hennessey 2010, online). This action would have a precipitous effect on the transition process. In an interview with the author, one former official gave his view that this decision was 'quite a game changer... you're then beginning to evolve a narrative that is looking at the sort of last lap scenario. Whereas Gordon Brown's more expansive view was thinking a long way into the future' (Personal Interview 2013). This process was formalised by Cameron at the NATO Lisbon Summit in November 2010, which set the end of 2014 as the deadline for complete transfer of security and combat operations to the Afghan National Security Forces (NATO 2010). The Foreign Affairs Committee took a dim view of Cameron's actions, commenting in February 2011 that,

> the Government's current national security narrative is out of step with the current situation and, in light of the announcement of 2015 as a date for combat withdrawal, now out of line with the general thrust of UK policy. The 2015 date jars with the Government's national security justification which signals something very different; namely that the UK must do whatever is necessary to secure the safety of British interests. The two positions are not compatible and send mixed messages to the public. (Foreign Affairs Committee 2011: 87)

In effect, the Lisbon settlement represented an agreement between partners within the collective security framework of NATO that reflected a loss of political will to continue stabilisation efforts at the level and for the duration of time stated as necessary in Obama and Brown's strategic reviews of just 18 months earlier. Britain's counter-terrorism policy narrative ostensibly supported this eventuality by gradually narrowing the scope of discourse, allowing the substance of Lisbon to fit the narrative rather than jarring with it, as would have been the case had Blair and Brown's strategic narrative not been deflated. The issue that the Foreign Affairs Committee highlighted, however, was that in reframing the counter-terrorism narrative as one concerned only with national security in order to maintain public support for seemingly indefinite operations in Afghanistan, it also allowed the Government the pretext for putting in place a definitive timetable for withdrawal at the expense of the collective security system's strategic requirements. This was nothing less than an irony of the most monumental proportions. The national security focus within the counter-terrorism narrative had been produced in order to put the expanding aims of the conflict in Afghanistan (and particularly in Helmand) back into perspective with the original purposes of the mission, but paradoxically only served to undermine the rationale. By lowering expectations of success to the point where it arguably contradicted the strategic requirements of national security, the counter-terrorist narrative had effectively out-rationalised itself. It had emerged from a situation where the political will for nation-building – the entire point of ISAF and stabilisation post the initial phase of counter-terrorism operations in 2001–2002 – was fast disappearing, and attempted to salvage the mission by reducing the importance placed on long-term stabilisation. It was a discourse couched in counter-terrorist rhetoric but with the referent object of a campaign that was fundamentally concerned with stabilisation; if the object was counter-terrorism then, as Cavanagh opined, bothering with human security issues appears in retrospect rather superfluous. The sum of the narrative configuration was the creation of a paradox for strategic communicators. The counter-terrorist narrative re-emerged as an attempt to effectively communicate the purpose of the stabilisation element of the conflict, but in doing so actually subverted the purpose of the conflict to meet the demands of the narrative because the narrative belied the transnational nature of collective security requirements. In attempting to communicate 'strategically' to the British public that Afghanistan was about their personal security, the counter-terrorist narrative sought to circumvent the complicated theoretical story of stabilisation and all its connotations of liberal norms, collective security, multinational cooperation and questions of the relationship between direct and indirect threats. The end result, however, was the creation of a 'discourse trap' or narrative 'blowback', where the national security element, once embarked upon, overtook and rhetorically negated the liberal-normative basis of the strategic importance of stabilisation (Michaels 2013).

Phase V: 2011–2014 – Drawdown and Narrative Blowback

The killing of Osama bin Laden in May 2011 likely served as a fillip to the by-then entrenched counter-terrorist narrative for Afghanistan. It allowed statesmen and women to draw a direct link between the distant origins and aims for the conflict by pointing out that the figurehead of the al-Qaeda movement had been dealt with. Obama used it to his advantage in his June 2011 'drawdown' speech:

> We're starting this drawdown from a position of strength. Al-Qaeda is under more pressure than at any time since 9/11. Together with the Pakistanis, we have taken out more than half of al-Qaeda's leadership. And thanks to our intelligence professionals and Special Forces, we killed Osama bin Laden, the only leader that al-Qaeda had ever known. This was a victory for all who have served since 9/11. (Obama 2011, online)

This victory narrative may have provided national catharsis for Americans and provided a semblance of closure to the near decade-long quest for retribution against the alleged orchestrator of the 9/11 attacks, but it was left unstated as to how killing bin Laden aided the strategic objective of stabilising Afghanistan. The Coalition Government, by contrast, was forthright in asserting a causal link of the significance of bin Laden's death. In a statement to the Commons in July 2011, Cameron noted that 'Osama bin Laden has been killed and al-Qaeda significantly weakened' and that 'British and international forces have driven al-Qaeda from its bases', which in his view justified 'entering a new phase in which the Afghan forces will do more of the fighting and patrolling', with the mission 'changing from "combat to support"' (Hansard 2011). By extrapolation, what Cameron's statement demonstrated was that focusing on counter-terrorism in the wake of bin Laden's death allowed his fellow statesmen and women to play down the intermediate and difficult stabilisation and counter-narcotic aims of the Afghan campaign from the second, third and fourth phases of the conflict. It is worth noting that the counter-terrorist aims Cameron alluded to in his Commons statement had all effectively been achieved in 2002, following the joint US-UK counter-terrorism mission codenamed Operation Anaconda; by contrast, the work of defeating the insurgency and stabilising Afghan society, the true test of ISAF operations, was paid short shrift. The reasoning behind this appears to be quite clear: during this final phase of the conflict, the counter-terrorist narrative underwent a process of reconfiguration to reflect what had been accomplished as ministers shifted emphasis towards 'drawdown' efforts in the country. This may be interpreted as Government taking on a new line of highlighting only security-related aspects, mainly in the form of noting progress in ANSF capacity building efforts and thereby downplaying any perceived failings of stabilisation by moving the goalposts of success.

As the drawdown process gathered pace through 2011 and 2012, the Coalition's counter-terrorism narrative was reined in to give an impression of a task completed. Interestingly, Cameron's Commons speech noted that while Gordon Brown 'told the House that some three-quarters of the most serious terrorist plots against Britain had links to Afghanistan and Pakistan', the proportional threat posed by the region to British national security was 'now significantly reduced' (Hansard 2011). Whether the threat was reduced in relation to 2009 when Brown frequently referenced that statistic or in relation to 2001 was not clear, nor was it evident by what metric or to what degree a threat must be reduced in order to justify an accelerated withdrawal. Whatever the case, members of Cabinet, most notably newly appointed Defence Secretary Philip Hammond, continued to make the case for British involvement in Afghanistan as being primarily about 'keeping us safe from terrorism on our streets' (Hammond 2011, online, 2012, online). Throughout the remaining three years of the drawdown process, such revisionism remained a staple part of the counter-terrorism narrative. To illustrate: in December 2011, Chief of the Defence Staff General Sir David Richards reminded his audience that

> our own national security underpins what we are doing in Afghanistan. Ten years ago I would have felt no need to mention it. It is interesting to note how quickly many outside government forget that the ungoverned, unstable space that was Afghanistan became everyone's problem on 9/11 and the UK's own home-grown 7/7 bombers were trained in Pakistan. (Richards 2011, online)

While Richards' comment on the tendency of the public to lose interest over time was undoubtedly accurate, the notion that 9/11 was a 'national security' issue rather than a collective security issue is highly problematic, as is the case for Afghanistan as a national security imperative when, in the next sentence, he admits that it was Pakistan, not Afghanistan, that accommodated the 'training' of the 7/7 bombers. In this way, Richards' comment demonstrates how the justificatory framework of collective security remains the only sensible and coherent way of explaining British participation in Afghanistan. A national security-based argument encounters problems such as those described by Richards: 9/11 was an attack on the United States (unless one directly appeals to the logic of collective security and mutual aid), and Pakistan is not Afghanistan (the latter of the two being the focus of British national security issues). Thus it would appear, as King (2011: 388) remarked, that the political and military elite of the United Kingdom remained perplexed in how best to communicate the transnational dilemma that has characterised British participation in Afghanistan. Rather, as Richards' comment indicates, uncomfortable truths and inconvenient facts are simply brushed aside, ignored, or worst of all, uncritically subsumed within the chosen narrative pathway. This is not a uniquely British practice, either; take, for example, the following comment by NATO Secretary General Anders Fogh Rasmussen:

> Never forget that the main reason why we are in Afghanistan is to prevent the country from once again becoming a safe haven for terrorists that could use Afghanistan as launching pad for attacks against Europe and North America and

in that respect we have been quite successful. You haven't seen major terrorist attacks launched against Europe and North America since we started the military operation in Afghanistan. (Rasmussen 2012)

Rasmussen refers to the international political objective of counter-terrorism but ignores the strategic objective of stabilising Afghanistan. Interestingly, the national security line taken by Britain and Rasmussen is, according to research by Jens Ringsmose and Berit Børgesen (2011: 521–3), strikingly similar to that of Rasmussen's Afghan narrative during his post as prime minister of Denmark – a country, incidentally, whose sacrifices to the Afghan campaign have exceeded those of Britain in proportion to its population. Rasmussen could argue that the political objectives underpinning his reliance on counter-terrorism had largely been achieved (even though the original aim under Bush and Blair was to eradicate terrorism as a method, not simply to remove 'safe havens' for those who practice it), but even by this minimal criteria he makes revisions by pointing out that there have not been attacks on the scale of 9/11 'since we started the military operation'. The point of stabilisation, of course, was to leave Afghanistan in a state where future attacks – planned and carried out after the withdrawal of NATO troops – are prevented. In emphasising the original purpose of the mission, which the public is frequently called upon to remember, it would appear that a corollary is the desire for the public to forget the non- counter-terrorist aims of the mission. Narrative revisionism proved, in short, the only way for politicians to assert a modicum of success to the mission; this thereby allowed Cameron to make the argument that 'what we have done in Afghanistan is we came here to stop it being used as a base for terrorist activities. That has been and is successful' (Watt 2013). Asserting success required the negation of stabilisation and counter-narcotics-related aspects of the campaign.

Conclusion

Counter-terrorism was the original and last policy narrative employed by British Governments in explaining the purpose of its participation in Afghanistan. It is obvious that the Afghan campaign began as a response to the terrorist attacks of September 11, 2001, and has continued since 2001 with the objectives of either eliminating terrorism as a method of warfare or of managing the threat of terrorism emanating from within the borders of Afghanistan. In the most minimal sense, this policy has been consistent throughout the 13 years of the conflict. However, in terms of the policy narrative – the sum total of statements, justifications, and rationales uttered by British officials over time – the articulation of the counter-terrorism approach to Afghanistan has changed markedly. When taken in tandem with the two other policy narratives employed by the British state explored in the previous chapters, it is clear that emphasis on counter-terrorism has not been particularly consistently delivered either in of itself or in relation to the other two narratives. Counter-terrorism began as a liberal normative and collective internationalist policy, and ended as a rationalist, nationally-oriented articulation of UK security interests. Counter-narcotics and stabilisation policies emerged out of counter-terrorism policy as strategies designed to deliver the ultimate objective of a pacified Afghan civil society largely freed from the spectre of international terrorism, but along the way these two ancillary policies became the dominant policy narratives utilised by Government. This chapter has shown how the counter-terrorism policy narrative effectively disappeared from state articulations explaining the significance and purpose of British participation in Afghanistan as a consequence of the increased focus on ancillary or supporting policies, and only re-emerged as the central focus of such communication efforts once it became overwhelmingly evident to state elites that the counter-narcotics and stabilisation policies were failing to succeed on the terms by which they were originally pursued. A major consequence of the restoration of the counter-terrorism policy narrative as the pre-eminent line of justification for Britain's continued presence in Afghanistan was the downplaying of ancillary objectives required to achieve this aim, thereby doing harm both to those objectives and to the internal logic of the counter-terrorism policy narrative as well.

A strong case can be made for explaining the ebbs and flows of the counter-terrorism policy narrative as a consequence of transnationalised security dynamics. Counter-terrorism policy in Britain began as a result of following the American policy approach and the fact that British operations in Afghanistan were ultimately beholden to US strategy. The Blair Government's approach to articulating its policy was, in essence, to frame its own activities in Afghanistan as closely as possible to those of the Americans. Again, this was not conceptually difficult for Government given its predilection towards maintaining the 'special relationship' and, more generally, given its affinity for framing national interests as normatively grounded in the doctrine of international community as articulated by Blair in his 1999 Chicago speech. This in turn allowed Blair to give a normative basis to counter-terrorism as mutually supportive of democratisation and stabilisation, and to frame the threat of terrorism as not a national issue, but rather one of global significance that defied the use of national specificities. Collective security and counter-terrorism were conceptually inseparable and rhetorically gainful for Blair; his totalising approach to foreign affairs thus afforded him the ability to see the ancillary policies of counter-narcotics and stabilisation as part of a coherent whole of strategy (where the British could plug the gaps the United States had ignored) and policy.

Centrally, however, the Blair Government's acquiescence to the United States meant that its counter-terrorism policy (as well as its ancillary policies of counter-narcotics and stabilisation) was almost entirely dependent on that of Washington, thereby rendering its strategies subservient to inter-state dynamics of policy rather than necessarily what was required on the ground. On the one hand, one may view the counter-terrorism narrative as less a result of the transnationalisation of defence policy (as has been the case with the stabilisation and counter-narcotics narratives), and more an example of how transnationalisation has been rhetorically rejected for reasons of domestic political expediency and of justifying a reduction in commitments to Afghanistan. Narrative frameworks prioritising liberal peace theory and a normative, institutionalist approach security possess a significant explanatory power for collective security operations. They explain the purpose of a conflict in international or transnational terms by appealing not to national interests but to the indivisibility of individual nations' interests. This is possible by framing the object of security as one of values, thereby transcending distinctions that otherwise might be made concerning the policies of individual states and the undoubted differences in levels of threats felt by various NATO member states. They place a high degree of importance on combatting illiberal state practices and the transformative power of democratisation as an end in itself and, crucially, as the only means by which security can be attained. In short, liberal norms bind collective security members together in how they understand themselves and how they relate to both the mission at hand and their allies; they have a distinctive and unrivalled constitutive power. Thus, in this reading it is possible to view a shift from such language to that of counter-terrorism – focused as it is on 'national' security and interests –as a rhetorical move away from transnationalised defence policy.

On the other hand, one may see counter-terrorism's current dominance in Government defence narratives as the most powerful and effective means available of maintaining transnational defence policy under conditions that are simply no longer amenable to a liberal normative explanatory framework. In the absence of high idealism to coordinate the interests and policies of collective security members, appeals to counter-terrorism as the foundational and inalienable rationale of collective security efforts in Afghanistan (and elsewhere) can be understood as an alternative (indeed, perhaps the only alternative) narrative pathway for maintaining alliance solidarity. Counter-terrorism does not meet any high idealist principles (this chapter has shown how it categorically rejected any such principles), but it does contain within its narrative structure a not inconsiderable 'high stakes' rhetoric that is highly useful in maintaining transnationalised defence policy. Terrorism is perceived as an 'existential' threat by many NATO member states: it threatens global security as well as national security. The threat posed to citizens and infrastructure is a real one, and one that is readily apprehended and generally accepted by electorates. Articulating the purpose of collective security operations along counter-terrorist lines allows for the maintenance of transnational defence policy whilst simultaneously providing officials with the rhetorical freedom to place transnational policy demands within a domestic setting. In an ironic rhetorical turn, the language of transnationalised policy has been abandoned seemingly in order to secure the continued existence of transnational policy. More specifically, transnational policy has evidently moved beyond liberal normative articulations as a consequence of Afghanistan and Iraq, and the language of counter-terrorism can be seen as reflecting a much-reduced strategic programme that speaks to the bare essentials of individual states' national security interests within a collective security apparatus that is currently averse to ground interventions.

A synthesis of these two perspectives could be that the counter-terrorist narrative grew out of a flagging, norms-based stabilisation narrative that, whilst providing the most complete and accurate representation of NATO's mission in Afghanistan, had been overtaken by events and was no longer fit for purpose. This work has argued that transnational policy has shaped Government strategy and narratives, and that those narratives are geared ultimately towards the maintenance of transnational policy. This is because maintenance of transnational policy is an end in itself. The stabilisation narrative can be understood as reflective of Britain's desire to satisfy American expectations in contributing to collective security operations by adhering to transnational policy norms, but the process of doing so has stretched the UK state because it obligated itself to stabilisation policies it could not carry out. This in turn produced conflicts within the state that undermined the credibility of Government and the Afghan mission, thereby threatening the unity of the Alliance system. Thus, the emergence of the counter-terrorism narrative may be seen as an attempt to salvage the mission and, by proxy, to insulate the collective security framework from internal rot. This perspective allows us to see how the tensions between departments of state caused by the strategic dilemmas of the stabilisation approach were ultimately rectified by Hutton in 2008 in his repositioning of the meta-narrative for Afghanistan as one of 'national security interest' rather than being intrinsically about Afghanistan. This allowed counter-terrorism discourse to unshackle itself from the fetters of state-building policy narratives and to focus instead on British national security. SC's institutional emergence within the Ministry of Defence during Hutton's tenure suggests that the re-emergence of the counter-terrorism policy narrative was an effort by the Defence establishment to entrench its own interests by utilising policy to affect strategy, insofar as the counter-terrorism narrative at this time served to circumscribe non-military activities and stem the concordant entropy in messaging.

The re-emergence of the counter-terrorism narrative can therefore be seen as holding functional qualities beyond the mere focus on improving public understanding of the purpose of Afghanistan. Indeed, its ultimate contribution was to inform, through diplomatic manoeuvres in Washington and Lisbon, a justificatory framework for the redirection of alliance-wide strategy on Afghanistan to one of definitive timetables for withdrawal and the severe curtailment of stabilisation strategy in the country. The problem remains, however, that stabilisation and counter-terrorism were, in both narrative terms and in the substance of Afghan strategy, incompatible. They represented two poles in a zero-sum game. Indeed, the point of the stabilisation narrative was always to emphasise the insufficiency of a counter-terrorist approach since, according to liberal normative thinking, counter-terrorist activity would fail to address root causes of conflict and, in all likelihood, exacerbate terrorism in the long run. Likewise, the emergence of counter-terrorist discourses has carried with it the implication (seemingly by design) that stabilisation objectives must be downplayed and to a considerable extent discarded in order to sustain a nominal degree of commitment to Afghanistan by NATO members. Thus, the counter-terrorism narrative can perhaps be best understood as an attempt to preserve Alliance unity via a transnationalised policy agenda in Afghanistan by, paradoxically, undermining the language of transnational policy and, in doing so, the significance of Alliance efforts in Afghanistan.

Chapter 7

Conclusion: The State of British Strategy and the Utility of Strategic Communication

Taken individually, the three narratives for the British experience in Afghanistan explored in this book demonstrate serious deficiencies in the crafting and maintenance of strategic communication. This work has shown that these narratives were not consistent over time and in each case suffered from almost total incompatibility when studied as a unitary whole from the start of military operations in 2001 to its cessation in 2014. Despite the existence of strong arguments (see Ledwidge 2011, 2013; Betz 2011) that British communication efforts in Afghanistan were inconsistent – the fact that 'strategic communication' protocols are now in place should be evidence enough that such a view prevailed in Whitehall as well as in scholarship – this consensus is not universal. Ringsmose and Børgesen (2011: 515), for example, have claimed that UK messaging on Afghanistan 'has from the very beginning argued in a clear and consistent manner that the purpose of the mission is to protect the security interests of the UK'. The problem with their analysis is that they do not critically analyse UK communication efforts in their own right, but rather employ a comparative approach between NATO members, wherein the UK apparently performs better than some of its allies. The first and foremost contribution this work has made is to challenge such assumptions by undertaking an in-depth analysis of the UK's track record on communicating the purpose of Afghanistan on its own merits. This work shows that each of the three narratives analysed contained their own internal inconsistencies that militated against unified and coherent exposition, resulting from the various domestically and internationally located pressures of statecraft, diplomacy and the uncertainties inherent in military conflict and state-building. The three policies were never compatible with one another: counter-narcotics efforts were inherently de-stabilising, effective stabilisation meant accommodating narcotics to a great extent, focusing on counter-terrorism implied a reduction in emphasis on stabilisation, and the link between narcotics and terrorist activities was never as undisputable as the early policy suggested. It is surely the case that what has been lacking in Britain's presentation of its policies and strategies for Afghanistan has always ultimately been the 'lack of a workable political objective' and, from that, a dearth of 'clarity about what we are "selling"' (Betz 2011: 629; Ledwidge 2011: 256). This is a point endorsed by British SC doctrine in no uncertain terms: '[s]trategic communication will fail where there is an absence of policy and strategy, or it is not crafted integral to the strategy, or what is said and done do not align with the strategy' (DCDC

2011: 1–4). Indeed, one may be forgiven for assuming that these words were written with Afghanistan in mind, for each of these shortcomings is evident in the empirical work that precedes this conclusion.

If one can speak of a 'meta-narrative' encompassing these three divergent policy strands, it is difficult to conclude that such incompatibility could have ever produced a coherent whole. When attempted, the meta-narrative under Blair can be seen as an attempt to coalesce incompatible sets of policies (and the interests behind them) into a consistent conglomeration despite overwhelming evidence suggesting otherwise. Thus, for the first half of Britain's involvement in Afghanistan, the policy narratives became detached from political and strategic realities. They began as a Kantian articulation of the importance of collective security norms and internationalism, human security and liberal interventionism, democratisation and comprehensive nation-building, claims for the innate existence of universal values and categorical imperatives, and an idealistic belief in the transformative power of a 'moral use of force'. They ended as a realist articulation of the centrality of national security and national interest and a downplaying of collective security logic, a minimalist conception of stabilisation that largely precluded non-security aspects, the relativity of values and a prioritisation of the British subject over all others, and a deep-rooted pessimism regarding the utility of the use of ground forces in democratising missions. When comparing the messaging of the first half of the conflict with that of the latter half, it is evident that there exists a state of binary opposition on practically every conceivable distinguishing feature of the meta-narrative. This point is made more interesting when one notes that the latter half of the meta-narrative appears in direct contradiction to the rhetorical and philosophical ethos of the Blair Government, which maintained that the meta-narrative was a coherent whole, that counter-terrorism, stabilisation and counter-narcotics not only could coincide with one another, but that the meta-narrative and overarching strategy for Afghanistan would not make sense without their coalescing. Whereas Blair may have believed that Britain is 'a country that could combine opposites and reconcile the contradictory', it was never apparent how his tendency toward narrative synthesis could work in strategic terms on the ground (Freedman 2007: 616).

The point that needs to be stressed here is that Blair's penchant for reconciling contradictions was, in this author's view, a reflection of his Government's proclivity for seeking to achieve the optimal presentation of reality. This was at times arguably more important than the reality itself, for obvious reasons of gaining advantage in the maelstrom of party political gamesmanship and the less obvious reasons of maintaining the UK's relevance amongst its international partners. Because the reality of things was always of less importance than the presentation of reality, a natural consequence was a dearth of strategic planning. Strategy, to reiterate, is the linking of ends (policy objectives) with ways and means (the tactical and operational methodologies employed and the resources at one's disposal to achieve those ends), and exists within a dynamic and highly contingent environment. Under Blair, the ways and means of achieving the ends of policy, and the dynamism and complexities of the situation faced in Afghanistan, were never fully apprehended primarily because they served as impediments to the realisation of the strongest possible articulation of the policy – i.e., of the strategic narrative pathway. This narrative pathway was attached to the policy of stabilisation as a means of demonstrating British commitment to Afghanistan, and through that its loyalty to the United States and the transatlantic alliance. In this line of inquiry, it is possible to view the original narrative for Afghanistan as remarkably strong, focusing as it did on a core idealist philosophy that encapsulated the political zeitgeist of the post-9/11 era: a bold and assertive foreign policy, a focus on internationalism and the combination of military force with a benign developmental agenda and, above all, a firm conviction that Britain could balance its self-interest for security, freedom and prosperity (via collective security and the practice of foreign internal defence) with an ostensibly benevolent programme of ameliorating global injustices and inequalities. Most importantly, Blair's worldview allowed the synthesis of seemingly contradictory aspects of statecraft – political realism and liberalism – by framing whatever the collective interest happened to be as fundamentally synonymous with the 'national' interest of the United Kingdom, simply because the national interest was declared to be only realisable within a collective setting. Such a philosophical position, while politically useful in that it was oriented directly at Britain's core interest of supporting the United States, also prevented the realisation of strategic prudence for the United Kingdom since it for the most part negated the conceptual basis of "national" strategy. Conversely, the ways and means of strategy, properly analysed and applied, would have mitigated the possibility of such an articulation and instead produced a rather staid and banal meta-narrative – much like the one that has operated since late 2008. The problem with this approach is precisely the opposite of that of Blair, however, in that it negates the possibility of a candid and open strategic re-think by obfuscating Britain's subservient role to the United States, a precept that remains at the heart of most considerations of defence policy and strategy.

Seen in the light of this dilemma, then, it is apparent that the task of SC practices was to bring political rhetoric back into line with political possibilities. In order to rehabilitate the core narrative (counter-terrorism), the ancillary ones (counter-narcotics and stabilisation) were subjected either to marginalisation or complete dismissal. Although central to the UK fulfilling its political obligations to the international community, the pursuit of stabilisation and counter-narcotics policies created severe institutional tensions within the state. As such, the only recourse for British officials to win back public, military and political support for the mission was to effectively reach back into the past to a point prior to the advent of the stabilisation and counter-narcotics narratives and re-present the meta-narrative for Afghanistan as being about – and as only ever have been about – counter-terrorism and British national security.

An implication made in this work is that the reconfiguration of the Afghan narrative undertaken since 2008 was one that was Defence-led, and as such primarily reflected defence interests. In a 2011 Chatham House report, Paul Cornish, Julian Lindley-French and Claire Yorke noted that, unlike in the case of the United States, which has developed its own version of strategic communication ostensibly as a pan-government initiative, British SC had, at the time of publication, been largely limited to defence activities. In observing that defence doctrine has preceded a cross-government initiative, Cornish et al. inferred potential deleterious consequences: 'the British approach has some of the appearance of the tail wagging the dog, in that until recently the defence contribution has been offered before a national concept has been established' (2011: 5). In other words, national strategic communication practice appears to be led by Defence primarily with Defence purposes in mind. This point has been confirmed by this work: the institutionalisation of SC within the MOD emerged out of a specific point in time during the Afghan campaign in which the meta-narrative's future trajectory was in the balance, between one that favoured a national security approach and one that remained focused on the liberal interventionist mantra of national security through development and the pursuit of human security ends. It is in this context that we may understand strategic communication as primarily about the pursuit of 'the national interest' along the narrative lines of defence and security. The empirical evidence supports the thesis that SC – in its domestic form – was ostensibly undertaken in the United Kingdom as a means of realigning the mission along the lines of its original purpose of counter-terrorism because of the strains exerted upon the liberal normative meta-narrative by the emergence of the transnational dilemma into public discourse. This appears to have been done in order to cement into the narrative the military aspects of the intervention and, as a corollary, to undermine the almost entropic tendency of non-security aspects of the meta-narrative to expand exponentially in response to the growing demands of the Comprehensive Approach. What all of this leads to is the conclusion that strategic communication as currently practiced within Whitehall, and particularly that which emanates from the MOD, is 'strategic' insofar as it is concerned with supporting (consciously or not) the interests of one element of the state – Defence – in the face of wider state apparatuses whose predisposition is aimed at democratising and developmental efforts. As a matter of course it rejects many of the core tenets of liberal internationalism and interventionism espoused by the Blair Government in the earlier stages of the Afghan campaign, rooted as it is in realist conceptions of national interest and security as the referent objects of British defence and foreign policy. This grounding means it cannot easily incorporate the use of powerful ontological assertions of universality or grand visions of the transformative power of liberal interventionism, nor can it decisively appeal to the principles of sacrifice and mutual aid central to collective security membership, since it is foundationally based on a discourse of self-interest. In short, strategic communication in its current form appears as an institutionalised safeguard against the excesses of liberal strategic logic.

Thus, in sociological terms, communication is strategic because it promotes state-wide adherence to a traditional, realist understanding of strategic thought that is grounded in the interests of the foundational unit of the nation-state, as opposed to a liberal international relations theory understanding of its basis in collective security mechanisms informed by interdependence dynamics or, beyond that, by human security discourses that place the referent object of security at the level of individual human beings. In doing so, SC goes some distance in resolving the transnational dilemma facing the United Kingdom outlined in the first chapter of this work by utilising policy narratives as a weapon of statecraft. Disagreements between alliance members over appropriate levels of 'burden-sharing', informed in part by divergences of political culture, policy objectives, and national strategies remain unresolved, and are unlikely to drastically improve in the short-term. This has created pressure for states such as the United Kingdom, which have borne a disproportionate amount of the collective burden, by making it more difficult to utilise narratives that emphasise the importance of engaging in collective security operations and the benefits of membership to a collective security community. In response, SC practices have reflected a natural tendency – in the face of the problem of free riding – of reasserting a realist understanding of interests and emphasises, in no uncertain terms, the importance of stabilising Afghanistan in terms of a singular, national security interest. A more proscribed and nationally-focused form of strategic communication combats the excesses of the values-centric rhetoric of liberal foreign policy that NATO adheres to, giving British policymakers relatively greater freedom in how they articulate the purpose of interventions to the British public in the future. This also allows for a kind of corrective function in the policy process by keeping political discourse and strategic narratives grounded in a realist appraisal of what is possible, and as such prevents the possibility of producing a narrative detached from reality that may potentially produce 'mission creep' and lead to intractable conflicts driven by unachievable objectives. This is what makes strategic communication 'strategic' in a classical sense: the linking and delimiting of available policy options and the speech acts that constitute and make real those options, to similarly delimited operational approaches used to achieve those policy ends.

National Interests, Transnationalisation and the Problems of Strategic Thought

The second contribution this work has made is an engagement with recent literature on UK strategy and the role of 'national interest' as its referent object. Throughout this book, I have argued that it is a fundamental lack of state autonomy over policy that has led to unworkable political objectives and a corresponding series of unsuitable strategies for Afghanistan. I have argued that Britain's defence policy is sufficiently subject to transnationalisation processes so as to make autonomous decision-making unviable, and that transnational policy tends to produce strategies that are highly contingent upon the institutional dynamics of those involved. The pursuit of such a policy has had an iterative effect on the British state, where political obligations and state transformation, undertaken to the ends of supporting collective mechanisms have led to severe intra-state tensions. Such developments have induced significant strains on the state as it continues to struggle to both define its role in the world and implement policies and strategies to meet transnational expectations. In the case of Afghanistan, where transnationalised policy has produced considerable political costs on successive Governments, one may observe the emergence of a 'transnational dilemma', wherein the obligations of the British state to its collective security partners compels it to undertake activities that, on the balance of costs and benefits, are difficult to conceptualise or explain as being in the national interest.

In this line of inquiry, one can view the existence of the 'selling' exercises or 'performances' of SC as a direct response to the structural and functional limitations of British policy and strategy processes as they become exposed against the challenges of collective security operations such as those described in this work. Conflicts require justifications, and conflicts without clear objectives (or with objectives unrelated or additional to stated objectives) require more persuasive narratives to justify them. Various ways of relating far-flung conflicts to the immediacy of the public's imagination have been deployed to this end, and most of these have centred upon connecting British operations in Afghanistan to a traditional, rationalist continuum of state activity that links interests to policies and policies to strategies by framing participation in Afghanistan as foundationally about protecting British citizens from Afghan-based or Afghan-inspired terrorists. Naturally, this articulation for Afghanistan jars with those made at the start of the campaign by Blair; with the existence of counter-narcotics and stabilisation policies in Afghanistan; and, of course, with the fact that Afghanistan can, in the main, only be deemed an existential threat to British national security by way of the United Kingdom's association with the United States. As has been established, however, all of these factors arose from a transnational diplomatic scenario wherein Britain's ability to autonomously and fundamentally re-shape the policies and strategies informing its narrative articulations would inevitably contradict its own core interests. In this sense, refining communication on Afghan policies was the only 'strategic' work that Britain could do, insofar as focusing on improving public opinion on Afghanistan appeared at certain points to be a greater priority for policymakers than re-evaluating the feasibility and purpose of the mission altogether.

This is not to claim that strategic communicators have necessarily behaved with duplicity with the British public about Afghanistan. Rather, it would appear – given the frequent commitments to internationalism given by successive Governments in national security strategies and strategic defence reviews – that the gulf between international diplomatic reality and British policy narratives is one that is deep-rooted in the social conventions and culture of the state itself. Part of my argument has centred upon offering an explanation for why it is that British Governments since the end of the Cold War have done so little to rectify the strategic shortcomings it faces. My answer relies on constructivist theory: specifically, I have argued that the obligation-cooperation cycle at the heart of the

UK's transnationalised defence policy has inculcated a situation where interests are constituted and replicated in part by the repeated act of pursuing those interests. When interests are pursued collectively, the pursuit of a common good has a normative constitutional force to it, allowing for the idea of international society where the pursuit of collective interests becomes a normative pursuit of its own, equal with and giving moral force to the pursuit of material interests. In this way, the iterative processes by which British interests are pursued have come to be partly constitutive of its national identity, curiously in a way that is fundamentally transnational in disposition. In other words, British state identity is in all practical senses one that perceives collective security and the special relationship as the normative and material bases for its own freedom, security and prosperity. Consequently, distinguishing between different sets of interest is not only difficult, it is also a question that is rarely asked, since collective security is taken to be so obviously beneficial to Britain that it is unimpeachable as a logical basis for its international standing. This is so for the simple reason that conceiving British foreign policy without it being a member of international society would be, rather paradoxically, to act against the national interest. For Britain to exercise significant agency in the world, it must work within a transnationalised defence structure that negates much of its agency to act independently.

This has not prevented the resurgence of 'national interest' as a referent object of strategy, however. Paul Newton, Paul Colley and Andrew Sharpe, for instance, argue that 'a clearly articulated Grand Strategy begs a plain definition of what is in the national interest' (2010: 48). Luis Simon and James Rogers (2011: 57) have similarly argued for a renewed understanding of UK interests on the basis that the United States 'has grown less able and willing to place British designs at the centre of its own effort', while Peter Layton (2012) has called for greater consideration of 'grand strategy' in UK policymaking. These studies stand in sharp distinction to those of Tim Edmunds (2011), Anthony King (2010) and Patrick Porter (2010). Porter in particular argues that the idea of British grand strategy is problematic precisely because the UK does not operate independently of the United States, but rather 'continues to organise itself as a satellite' of American strategy (2010: 9). Indeed, from a constructivist perspective that accounts for the iterative effects of transnationalised defence policy articulated by Edmunds and King, advocates of a 'return of British strategy' evince a kind of realist 'black boxing' of state agency. They point to 'national interest' as the referent object of strategy but take it as a given rather than as a term requiring deeper analysis. Where such analysis is provided by realists in asserting the term's use for defence policy, as Colin Gray has done, it takes the form of a set of ordered priorities of interests crucial to national "survival", those that are "vital", those that are "major", and "others" (2010: 169). The transnational structure that pervades UK strategic and defence policy is not present in such discussions, however; this work demonstrates that transnational demands may influence the prioritisation of defence policies more than any assessment of their intrinsic utility to the British state or commonweal.

Indeed, this unproblematic positing of national interests goes hand-in-hand with a realist perspective precisely because the existence of clear and unambiguous national interests is what underwrites the activity of strategy itself. For realists, national agency (as opposed to transnational role-playing) requires national strategy, which must be based upon national interests. Maintaining the possibility of national and grand strategy 'privileges the power of agency' (Suhrke 2013: 272), whereas imputing the pervasiveness of transnationalised policy's structural limitations on ideas of strategy and interest implies that 'one is essentially admitting loss of governmental control' over its own agency (Eriksson and Rhinard 2009: 253). In other words, once the 'national' element of interests is problematised, the potential for national or grand strategy to exist – much less to solve the problems of current UK strategy or the lack thereof – becomes less clear. It is for this reason that I have argued that the utility of 'national interest' is largely limited the discursive arena – that is, to dealing with the political drawbacks of transnationalised defence policy. Strategic communication exists within this paradigm, so I have argued, to serve the purpose of preserving British interests, its concordant policy positions, and the illusion of national ownership of strategy against its own logical inconsistencies, all of which stem back to overcoming the difficulties of conceptualising and explaining transnationalised policy to domestic audiences. Its functions include unifying the state through a narrated reality of 'national' interest, which promotes the alignment of various interests within the state and thereby acts as an institutional measure aimed at the avoidance of information fratricide; utilising such processes as a means of gaining public support for collective security by framing operations as nationally based – in other words, to protect British national interests secured via collective security mechanisms by not mentioning those mechanisms; and overcoming a lack of control over policy and strategy by using 'strategic' communication of policy to influence the behaviour of external state actors by signalling its unwillingness to tolerate imbalanced burden-sharing – otherwise known as the 'defence' approach to communication.

The result of all three of these functions is that policy has increasingly become synonymous with strategy within the British state. This is because the maintenance of British interests via transnationalisation reduces the state's autonomous agency, and requires policies that produce strategies (in this case, the 'strategy' of communication) that do little more than refine and justify the discursive delivery of policy in order to present British interests without reference to transnational dilemmas. This has the strategic purpose of influencing public opinion that British 'national' interests are at stake (and that is not to say they necessarily are not), thereby securing support for a given policy and positing the compatibility of transnationalised policy with traditional understandings of national security. In this sense, communication can indeed be strategic. Moreover, by the metric of Britain's national interest in maintaining collective security mechanisms, SC has been rather successful: Britain's relationship with the United States remains strong and NATO remains intact. Herein lies the problem for British statecraft, however:

the equation of policy with strategy, and of communication as a kind of strategic policy or politicised strategy, means that the search for a solution to the state of British strategy is simply ignored. The rise of SC indicates the demise of strategy: in the absence of meaningful control over strategy or policy for Afghanistan, the work of SC has been one of realigning complex transnational reality with a simplistic, politically advantageous story. Rather than coming to terms with the strategic shortcomings surrounding Britain's role in Afghanistan, the trend in political communications has been simply to posit an alternative reality where failings were not really failings because they were never really important to the political and strategic objectives of the mission in the first place.

Thus, whatever strategic communication's uses in favourably altering the presentation of Britain's strategic dilemmas, those dilemmas still exist; they will not simply disappear because states have devised new ways of talking about them. Reconfiguring Afghanistan as a 'national security interest' did not change the fact that the collective security through NATO remains the *modus operandi* of British defence policy, nor did it have a significant impact on inducing change within the political cultures of Britain's NATO counterparts to take on greater burdens in the conflict. Britain remains tied into frameworks that will demand sacrifices from the realm in terms of blood and treasure, and more often than not these conflicts will not be irrefutably and obviously related to an existential struggle for national survival (Porter 2010: 356). Rather, they will likely be precisely the type of risk management scenarios represented by stabilisation missions such as that in Afghanistan, Iraq, and more recently in Mali, or in more limited engagements such as in Libya and the campaign against the 'Islamic State' in the Levant. Thus, in the face of such institutional obligations, it is clear that merely talking about British defence policy in realist terms cannot overcome the inherently liberal internationalist posture of UK foreign relations; the point, of course, is that because this is the case, it is also evident that policymakers do not actually aim to transform the substantive content of British defence policy into an avowedly realist formulation.

There is, therefore, a communication gap at the heart of British statecraft between what must be said – that all interventions undertaken in the name of collective security are in fact directly in the national interest, for national security reasons, or in the 'national security interest' of the United Kingdom – and what cannot be said – that even if they are not *directly* in such interests, it behoves the state to participate regardless. This is a double-edged sword, of course. The former is often a convenient truth since, because the British 'national interest' is logically inseparable from the collective security interest, all collective interests may be understood as national interests. The latter, however, is an inadmissible reality for the same reason: even though, in realist terminology, one may judge interests in direct and indirect levels or hierarchies regarding their importance to an individual nation-state, collective security imperatives do not allow this, for if they did alliance members would simply pick and choose when, where and how they wished to contribute to collective security operations and thereby compromise the effectiveness of the mechanism. This is, of course, the story of NATO in Afghanistan, and is an irony given that a major aim of NATO in Afghanistan was to strengthen the collective security mechanism. This also goes some distance in explaining Britain's repeated willingness to take on a disproportionate level of responsibility within NATO operations in Afghanistan, for it is the only means by which Britain's global interests can be maintained. Such involvement has come at a great price to Britain in terms of casualties, financial cost and its geopolitical and martial reputation, and as such has stretched the fabric of the UK's capability to explain its involvement in Afghanistan within the normative and material confines of collective security logic. Strategic communication will continue to be necessary in the future so that states such as the United Kingdom can articulate to their general publics the complex yet mundane workings of collective security mechanisms in simple and stark realist phraseologies of security, interests, and 'security interests'.

Given this analysis, it is difficult to avoid the conclusion that, in talking about collective security in strictly realist terms, British officials have contributed, consciously or not, to a distortion of reality. It is a simplification to claim that the primary reason for participating in Afghanistan was a response to a mortal threat to the security of citizens on British streets, just as it was a simplification to retroactively redact much of the narrative content of democratisation, development and counter-narcotics and posit, in its place, that the aim of British involvement was limited to counter-terrorism only. Even if the heightened emphasis on British national security and counter-terrorism was the 'whole truth' (and, as a caveat, it may well be, but the public – including, of course, this author – are not privy to the security services' unpublished intelligence estimates), the historical evidence of the narratives analysed in this work, as well as the evidence provided by Government regarding the 'chain of terror' from Britain to the tribal regions along the Afghan-Pakistani border, suggest not only the relative unimportance of the national security argument until 2008, but also that the primary threat to the United Kingdom in south Asia comes from Pakistan, a country with greater links to the UK and without substantial British military presence (and likely destabilised further as a result of NATO operations in Afghanistan). Reframing the Afghan narrative in national security terms was politically convenient, but by framing Afghanistan as a vital national security issue, rather than as an issue primarily related to Britain's strategic positioning within the international order, the state has precluded the opportunity for a reasoned public debate about the virtues of intervention there and about Britain's future role in the world. In the short term, it would appear that reconfiguring the Afghan narrative from a position of development, democratisation and global security to one of military activity, circumscribed stabilisation and national security paid dividends, insofar as it allowed a subsequent reframing of strategic priorities at Lisbon and facilitated a more rapid drawdown from the conflict. The longer term trends may be counter-productive, however. There is a danger that narrow realism will contribute to a growing isolationism in British political culture. Blair's aim to make a nation of 'reluctant global citizens' more confident and assertive to that end appears to have foundered, not least because of his own government's actions on Iraq. Almost certainly as a consequence of that legacy, British politicians today rarely speak in the grandiose fashion of Blair and Miliband, and instead opt for a language of prudence and caution regarding intervention. However, by delimiting available narrative pathways – even if for rational and sensible reasons – to the language of political realism, the British state risks creating a new 'discourse trap' where any advocacy of liberal interventionism effectively becomes viewed as an unthinkable and foolish policy option. This is what appears to have happened in the recent case of the failed Commons vote on Syria in August 2013.

In that case, Prime Minister Cameron tabled a motion calling on the House to agree that the alleged use of chemical weapons by the Syrian regime of Bashar al-Assad required 'a strong humanitarian response' and to agree that 'this may, if necessary, require military action' by the United Kingdom and its NATO allies (Hansard 2013a). Interestingly, whilst clearly making the case for the importance agreeing in principle to intervention in Syria primarily on normative grounds of maintaining the 'taboo' on the use of chemical weapons, and secondarily on humanitarian grounds and with the support and legitimisation of the international community, Cameron chose to emphasise these points by arguing that intervention was necessary for the national interest. While he refrained from positing, as one of his party's junior Members did, that Syria's use of such weapons constituted a direct threat to British national security (Hansard 2013b), Cameron argued that 'a stable middle east is in the national interest, but there is a specific national interest relating to the use of chemical weapons and preventing its escalation' (Hansard 2013a). While he received little argument on that point, the majority of his contemporaries – including over 40 Conservative and Liberal Democrat Members – disagreed with his premise that intervention in the conflict in Syria would serve the national interest, with the consequence that the motion failed. Indeed, during the debate some within the Conservative Party argued that maintaining the taboo on the use of chemical weapons was not a distinctively national issue and was, therefore, not a national interest at all:

> [w]hy is it any of our business? Has Syria ever been a colony? Has it ever been in our sphere of interest? Has it ever posed the remotest threat to the British people? Our job in Parliament is to look after our own people. Our economy is not in very good shape. Neither are our social services, schools or hospitals. It is our job to think about problems here. (Hansard 2013d)

The Government's failure to carry the motion raises familiar questions about its ability to communicate the linkages between the national and international and to relate the indirect requirements of collective security membership to the more conceptually simple terrain of direct threats to national security and 'national' interests. Displaying political will in opposing and potentially intervening against al-Assad was not a *national* interest even though it was in the 'national interest';

it was not a '*national*' security issue but it was important in terms of maintaining international order through which national security is sustained. Agreeing in principle to a willingness to use military force to that end was a collective security requirement and, given Britain's conception of its national interest as located within the international community, reaching such an agreement on the appropriateness of the use of force should have been relatively unproblematic. This explains Cameron's resort to the use of realist language in advocating such a stance, but in framing participation in collective action in these terms, it also explains how he provided those opposed to such action the discursive upper hand. Appealing to national self-interest permitted the ascendancy of a narrow, traditionalist and arguably isolationist articulation of political realism to come to the fore and compromise the Government's argument. In contrast, a Blairite appeal to human security and humanitarian and liberal norms and values, and the importance of collective security mechanisms in maintaining international order, would possibly have incurred less semantic confusion and contradiction. Interestingly, Cameron's former Defence Secretary, Liam Fox, for so long the architect of moving the Afghan narrative away from such arguments, opined in this debate along liberal interventionist lines in claiming that, while 'there is no national interest for the United Kingdom in taking a side in that civil war', Britain should show willingness to intervene in Syria for reasons of humanitarianism and global security (Hansard 2013c). This appears to be an archetypal case of discursive 'blowback': in utilising one narrative framework to escape the discourse trap of another in order to facilitate the extrication of forces from one conflict, the British state unwittingly created the conditions for the development of yet another discourse trap that would prevent the possibility of deploying forces in potential conflicts in the future.

Indeed, the pitfalls of utilising rationalist narratives to justify collective security operations continue to reverberate to the present day in reference to the situation in Iraq and Syria regarding the so-called 'Islamic State in Iraq and the Levant' (ISIL). Britain has committed itself to supporting an ad-hoc coalition of states led by the US and has undertaken air strikes against several ISIL targets in Iraq, but has explicitly ruled out the possibility of deploying ground forces in the region. Perhaps as a reflection of satisfaction with this limited level of engagement, the Commons overwhelmingly approved the use of force against ISIL. What is of greatest interest to this work, however, is the manner of the appeal made by Cameron to the Commons in making the Government's case for intervention. As with the failed motion of 2013 regarding Syrian chemical weapons, Cameron placed the instability of Iraq and Syria firmly within a rationalist discourse of national interest. Unlike the 2013 motion, however, the motion for intervening in Iraq in 2014 was based on the positing of a direct security threat to the United Kingdom. Interestingly, these two elements – national interest and national security – formed the basis of his argument:

If we are to do this, a series of questions must be answered. Is this in our national interest? In particular, is there a direct threat to the British people? ... The answer is yes. ISIL has already murdered one British hostage and is threatening the lives of two more. The first ISIL-inspired terrorist acts in Europe have already taken place, with, for instance, the attack on the Jewish museum in Brussels. Security services have disrupted six other known plots in Europe, as well as foiling a terrorist attack in Australia aimed at civilians, including British and American tourists... This is not a threat on the far side of the world; left unchecked, we will face a terrorist caliphate on the shores of the Mediterranean and bordering a NATO member, with a declared and proven determination to attack our country and our people. (Hansard 2014)

Again, it is worth reiterating the caveat that ISIL may indeed represent a direct threat to the UK in a way that would justify the apocalyptic quality of Cameron's statement: the details of intelligence at this stage are naturally not available to the general public. It is interesting to note, however, that Cameron makes the argument that ISIL are a 'direct' threat to the British people without providing evidence to substantiate that any direct threats to the British people have indeed taken place. Rather, he cites the death of a British journalist in Syria, an attack in Belgium, plots across Europe, and a foiled plot in Australia. None of these can be considered incontrovertible evidence of a direct threat, but more as constituting an indirect threat or a direct risk. Indeed, what Cameron appears to be speaking to is the principle of collective security of alliance partners rather than an issue of direct national security. The notion that Government has locked itself into a rationalist discourse to explain what remains a fundamentally transnational defence policy is further highlighted by comparing Cameron's rationale with that of Obama: just a fortnight before the Commons motion he declared, contrary to Cameron, that the US had 'not yet detected specific plotting against our homeland', but rather that 'ISIL leaders have threatened America and our allies' (The White House 2014, online). Again, the point is not to cast doubt on the veracity of their claims, but rather to note that, while Obama's rhetoric makes concessions for the level of threat posed to American citizens, Cameron explicitly refers to the direct and present threat to national security posed by ISIL rather than noting the empirically stronger case that ISIL had some involvement in attacks and plots against Britain's allies. Relying on realist justifications – particularly those that emphasise direct threats – appear to have become the only way Britain can articulate its collective security obligations to its domestic audiences.

The Dangers of 'Narrative-Led' Strategy or 'Strategising' Policy

Britain's transnational dilemma in Afghanistan has been one where the state has been subjected to conditions of collective security membership that have increased tensions within the state to find ways to extricate itself from the more unrealistic obligations it had in the conflict, but to do so in a manner that did not compromise its international relationships and standing. SC can be seen as the state's solution to this problem and as a solution that arose in institutional terms because of the existence of this problem. Reconfiguring narratives to achieve such ends between states is one matter; what has subsequently developed, however, is the academic and doctrinal supposition that these narratives have a 'strategic' value in shaping outcomes in theatre to meet the demands of coalition policy. The final contribution of this work is to query the validity of such postulations, given the ostensibly unintended consequences strategic narratives have had in the war in Afghanistan. It is telling that the majority of the mentions of 'narratives' in official British documents referred to those of the enemy, particularly that of al-Qaeda, and the need for a UK counter-narrative to fight a 'battle of the narrative' (Cabinet Office 2008: 26, 2010a: 16, 2010b: 6). However, over time the need for a counter-narrative has morphed into a more fundamental examination of the relationship between strategy and communication, lending itself to the development of strategic narratives that stand alone, independent of an enemy narrative, that describe strategy at the highest level. An example of this can be found in the current doctrine for British SC and narrative formation, *Joint Doctrine Note (JDN) 1/12*, which explains the need for a strategic narrative and communication as

> integral to strategy, both informing and supporting policy. Good strategy is usually forged from a single big idea, or a coherent collection of smaller ideas, with a clear underpinning rationale and unifying purpose. To be effective, the strategy must be instantly communicable if it is to gain traction at home and abroad. The logic of the strategy and its appeal should be compelling and easily understood. It must seek to gain and maintain the initiative and be set firmly in the context of the political purpose. It should bind the key players and the instruments of power, and in its totality should be simple, or at least capable of explanation in simple terms. (DCDC 2012: 1–4)

This excerpt demonstrates the vastness of the role assigned to strategic communication, including the remarkable contention that the 'logic' and 'appeal' of a strategy should be 'compelling' and 'easily understood'. Of course, if there is one aspect of collective security operations such as those carried out by Britain in Afghanistan that should be abundantly clear, it is that there is little about the strategies and policies therein that evinces compelling and readily apprehensible logic. Collective security operations by their very structure and function confound easy explanation – hence the need for SC and narrative, not just for the benefit of public understanding, but also so that policymakers tasked with prosecuting policy and strategy may understand what it is they are trying to achieve. Strategic communication exists because contemporary strategy is not easily explainable: the ability of a strategic narrative to explain a strategy is contingent upon the relative merits of the strategy itself, yet the need for strategic narrative and communication could only emerge from a scenario in which communication of strategy had been hitherto lacking in coherence and effectiveness, ostensibly because of a flawed (or ulterior) policy. Beyond the level of policy formulation, however, there is the issue of interest and collective security dynamics, the inherent complexity of which prevents the reconciliation of 'simple' explanations with accurate explanations. Perhaps this quandary explains why SC's role in 'informing' policy, and thereby guiding strategy, has been put forth as a potential solution to this dilemma. If the West is involved in 'a war of ideas', and its ability to explain a strategy is found wanting in a strategic information environment that places a premium on making strategic logic 'compelling and easily understood' to audiences, reason might dictate a strategy that is led or 'informed' by narrative. This notion is emergent in military and academic circles, with some even calling for the further development and institutionalisation of 'narrative led operations' (Nissen 2012). Carsten Roenfeldt (2011: 58) offers a paradigm of 'productive war' – based on force of arguments rather than force of arms as more relevant than traditional physical warfare since, in his view it is 'discursive clashes that determine political projects' core values'. By this, it is conceived that in the process of developing a strategic narrative, a state demarcates the values and principles that inform its policy and dictate its strategic options; as Roenfeldt claims, this is possible by flipping the Clausewitzian dictum of war being 'the continuation of politics by other means' on its head (2011: 53).

Emile Simpson's *War From the Ground Up* (2012) takes this inversion of Clausewitzian strategy as his starting point of inquiry, and is perhaps the most ambitious and well-received attempt at promoting the role of strategic narrative in contemporary conflict. In a manner similar to Roenfeldt, he claims that politics is now an extension of war, or more precisely that the realms of politics and war are fusing so that the centre of gravity of conflict tends to be primarily ideational, thereby making the enemy and neutral parties' acceptance of one's strategic narrative the primary goal in irregular conflict (2012: 75). As such, Simpson contends that successful strategy hinges upon convincing strategic narratives – essentially the articulation of policy, or 'policy narratives', which for liberal states involved in long counterinsurgency campaigns such as Afghanistan means relying upon advancing moral arguments to gain support. Such moral arguments, or 'ethos', are for Simpson informed by viewing 'vision and confidence in one's values' as 'the core of strategy' (2012: 221). In this way, Simpson argues that strategy in conflicts such as Afghanistan should be led or guided by a narrative that is centred upon the assertion of the ethical and moral rightness of the mission. This is challenging, however, since it is clear that a range of events and interests (the 'fog of war') militate against the coherent and consistent realisation of moral and ethical activity in pursuit of moral and ethical ends. This in turn produces the potential for contradiction within narrative. Simpson comes close on several occasions to identifying disparities and conflicts of interest as the primary cause of disjointed policy and, therefore, of untenable strategy:

> [s]trategy requires an abstract starting point, an idea, which is typically understood as policy. Clausewitz stated that 'policy is nothing in itself ... we can only treat policy as representative of all the interests of the community'. In reality, to reconcile all the interests of the community in relation to a given conflict is often impossible. Even those constituencies who support a military action may well do so for different, and potentially contradictory motives. The further one moves away from wars fought for national survival, the more likely is one to detect such inconsistencies. (Simpson 2012: 117)

This is an astute analysis that has been confirmed by the empirical chapters of this work. Policy is likely often representative of the interests of all involved in the policymaking process. This was how the comprehensive approach to stabilisation was devised, and how it was conceived to function. Of course, the irony of this is that narratives for Afghanistan failed to coalesce because of this scenario, leading to the failure of the comprehensive approach as a strategy and the eventual development of strategic communication. The issue for Simpson, however, is that he assumes that policy is 'the starting point' on a continuum that moves through strategy, operations and tactics (2012: 118). This is strange given his admission in the quotation above that policies arise out of interests, a point that should lead one to examine in depth first the manner in which policies are shaped by interests and how those interests inform strategies, rather than simply positing policy as a theoretical starting point of analysis. Because he neglects this issue, much of the potential for sociological and political examination of precisely the type of motives Simpson alluded to (those which this work has explored) is avoided, and as a result the analysis that remains – a largely abstract and hermetically sealed examination of the potential uses of narrative to the strategic process – is devoid of any consideration of the detrimental aspects of strategic narratives to ideas of strategy, much less to ideas of a strategic process guided by narrative. He does not problematise narrative; he starts with an assumption of it being a solution to strategy rather than as a symptom of a lack of coherent strategy. Such a perspective is natural in circumstances where the wider intellectual, ideological, political and social milieus directing state interests and policies to conceive strategy in a certain way – for instance, in a way that argues strategy to be essentially synonymous with articulations of policy – are left unaccounted for or are simply assumed to be a certain way. Of course, such a position leads us back to the starting point of this work and the subject of the problematic state of British strategic thought, one which Hew Strachan identified as such because of a recurrent incapability of those concerned to distinguish between policy and strategy (2005: 34,50). Hew Strachan's argument for improving this situation was similar to that which animated the MOD to undertake SC: it was based on a need for greater concord in the civil-military relationship (2006: 76–7). What distinguishes my argument, however, is the idea that fractures within the strategic milieu of civil-military relations arise from a discrepancy of the location of interests between the two parties: policy and strategy are not iterative or collaborative in a sealed binary coupling; rather, national policy is infused by external considerations that shape and direct its interests, which then cause discrepancies between what is strategically possible and what is politically necessary.

Of course, the sources of policy for any state, but particularly for the United Kingdom, are always multiple, taking in all the vicissitudes of inter- and intra-state compromise and negotiation as well as the personal or collective passions – be they ideologically or materially located – that animate policymakers in the decision-making process. Of primary importance, however, is a recognition that policy for the United Kingdom is substantially dictated by the collective security framework in which it operates, because it is on this basis – and only on this basis – that the United Kingdom may secure its 'national' interests – the rational referent of policy. This point is simply not recognised by Simpson or other contributors to SC literature, however; Simpson opts for a discussion of a Clausewitzian understanding of policy as 'in the final analysis, dependent on passion' (2012: 238). A possible result of neglecting this aspect of statecraft is that the majority of his analysis of the problems of contemporary strategy focuses on the misapplication of Clausewitzian paradigms of inter-state warfare in irregular conflicts; that is, he focuses on the root of strategic malaise as being located somewhere between policy and strategy, rather than as between the nodes of interest and policy. Ostensibly, this is because Simpson's investigation is one that moves 'from the ground up', applying tactical and operational lessons from his experience as a military officer to the level of strategy and policy. He extrapolates lessons at those levels to the level of strategy because he equates the power of narrative at the operational level – for instance, providing local villagers of a culturally genial story of the presence of foreign troops as a means of pacifying the area for stabilisation operations – to a potential power for narrative at a strategic level where persuasion is the centre of gravity and the performance of stories aimed at securing popular consent can become a paradigm for contemporary warfare. This is unfortunate in that its equation of the operational level of war with the strategic is a common mistake of proponents of the 'wars among the people' thesis. This work, by contrast, takes the opposite approach of policy from the top down, and thereby imputes that the majority of Britain's strategic shortcomings are not really strategic issues at all, but stem from intractable political issues at the transnational level which then exert themselves upon social dynamics at the strategic level of states in times of crisis. Simpson's is a common refrain from advocates of strategic communication and narrative, likely for similar reasons: they see the problem of strategy at the level of strategy and operations, rather than at the level of interests and policy.

To further illustrate, according to Simpson, the West's 'fixation' with 'generic doctrinal categories' such as 'insurgent' and 'government' produce the effect of 'upscaling' elements on the statecraft continuum: 'when operational ideas, which demand a political context, are not adequately provided with one they move up to fill the vacuum [of strategy or policy]' (2012: 228). Ironically, although voiced as a warning against the encroachment of politics into the domain of war, this is precisely what Simpson advocates in the form of strategic communication and narrative: the elevation of an operational accompaniment (which is in fact merely an articulation of policy) to the level of strategy. This approach appears, time and again, to result in a lack of discussion about the way in which conflicting interests can distort strategy and produce incompatible narratives, or where the fixation on the production of coherent narratives actually distorts the work of strategy by equating it with articulations of policy. The danger here is that what Simpson refers to as 'strategic narratives' are actually little more than operationalised versions of the 'policy narratives' investigated in this work. Indeed, the lack of such analysis can also be found in the tendency to prognosticate on the options for improving British 'national strategy' without even contemplating whether British strategy is 'national' and whether, given the demands of collective security on Britain and the absence of a clearly defined and independent national interest, a *national* strategy – or even strategy itself, insofar as it pertains to being both the means to an end (alliance cohesion) and the means to the means of that end (stabilisation missions) – is possible.

Again, without a critical understanding of the nature of interest, from which all things policy and strategy descend, it is much easier to develop a theoretical construct that looks at strategic narratives as relatively unproblematic or even supportive of strategy (as opposed to being indicative of a lack of strategy). Indeed, as mentioned earlier in this work, nearly all academic contributions on the subject carry an *a priori* assumption that strategic communication and strategic narrative are naturally beneficial to strategy processes, as opposed to one that sees them as existing as a consequence of a lack of strategy. An appreciation of the lack of strategy or a confusion of policy and strategy in Afghanistan might explain why the idea of 'strategic narratives' has apparently gained so much currency in analyses of the conflict by Simpson, Roenfeldt and practitioners like Stanley McChrystal. Few scholars have investigated the potential for 'narrative-led' approaches to distort and even undermine the strategic process as a result of a muddled understanding of interest, policy and strategy, and how these elements inter-relate, much less the idea that SC is the result of an already distorted strategic and political milieu. One exception is George Dimitriu (2012: 205), who notes how such a perspective could lead to unrealistic expectations on the power of narrative and communication to transform the state of contemporary strategy:

> There is the danger that too much value could be attached to Stratcom and that it could become seen as a corrective panacea against faulty policy or misconceived actions. Stratcom is in the end no substitute for an overall strategy, it is merely a supporting process. Not even the best communication and influence strategies are able to counter an unpopular policy or problematic actions.

This is a point well worth making. Strategic narratives tend to be given a level of significance to the policy and strategy process that does not take account of empirical realities. The received wisdom evinced in Roenfeldt and Simpson's work that posits a reversal of the relationship between politics and war as a reflection of changing contextual circumstances also assumes that the interests that underpin a traditional realist conception of power have been fundamentally altered. Moreover, the tendency of those advocating 'narrative-led' approaches seem caught in a paradox in which narrative is necessary because of contemporary uncertainty and chaos in international relations, but also impossible because – *obviously* – events dictate strategy, not the other way around. The idea that events can adhere to a script seems to contradict one of the main lessons of the War in Afghanistan (and indeed of much of the ongoing fallout of the Arab Spring), that is, the speed at which events can overwhelm understanding. Warning against strategic communication and narratives as a panacea to problems of strategy is therefore crucial. One can do worse in this regard than to recall Clausewitz's counsel on war as 'the province of chance':

> [from] this continual interposition of chance, the actor in War constantly finds things different from his expectations; and this cannot fail to have an influence on his plans, or at least on the presumptions connected with these plans. If this influence is so great as to render the predetermined plan completely nugatory, then, as a rule, a new one must be substituted in its place; but at the moment the necessary data are often wanting for this, because in the course of action circumstances press for immediate decision, and allow no time to look about for fresh data, often not enough for mature consideration. (1982: 140–41)

The fundamental paradox for strategic communication is this: in the pursuit of strategies of risk management, states have a natural tendency to try to mitigate risk – through the use of narratives that aim to support the pursuit of risk management by sustaining political support – but in doing so they actually produce the potential for creating greater risk to the political sustainability of risk management operations by attempting to predict, control and manage the course of those operations in defiance of a major philosophical tenet of strategy: that war is pervaded by elements of chance, contingency and uncertainty that defy attempts at prediction, control and management. In one sense it would appear that Simpson recognises this paradox by noting, in relation to the United Kingdom's 2010 National Security Strategy, that

> To consider broad global trends is no doubt of value. However, any expectations situated so far into an abstract future should at least be tempered by today's concerns...A distant-horizon gazing approach to strategy can leave one reacting to distant and fragile shadows that may vanish as soon as one approaches them. (2012: 242)

On the other hand, however, Simpson appears to contradict his own argument by promoting the development of strategies based on 'ethos', or the advocacy of moral justifications for intervention in stating that 'a strategic narrative which neglects ethos completely is in danger of finding itself illegitimate in the longer term' (2012: 213). Of course, this is an important point that has been borne out in the empirical chapters of this work: the production of the national security-centric counter-terrorism narrative at the expense of the normative interventionist strand that preceded it has likely come at the cost of nullifying much of the moral element in arguments for liberal intervention going forward. However, his view of the importance of ethos also stands in stark opposition to another major conclusion derived from this work, that is, that an adherence to an ethical and moral dimension to British stabilisation efforts in Afghanistan was precisely the reason for the strategic and narrative drift that ultimately brought about a reconfiguration of narrative (via the institutionalisation of SC processes) to one focused on narrow self-interest accompanied by the (often explicit) abrogation by public officials of the existence of moral or ethical elements to the strategic narrative. The point is that, just as strategic designs and policies vacillate in relation to events, so too does strategic narrative. Unlike in traditional narratives, strategic narratives are highly dynamic and unpredictable; they are not pre-destined stories waiting to be played out according to an already-written script. There is never one narrator, and the story being told is subject to forces beyond the narrator's control, thereby producing the potential for the entire narrative to be subverted and actually become detrimental to the policies and strategies it is designed to support.

A balance should therefore be struck in determining the utility of strategic communication and narratives. Discourse traps have been created by both the implementation of inappropriate strategic narratives and the attempt, through SC practices, to rectify the gap between rhetoric and reality. If one accepts that discourse traps have indeed occurred, then one must also accept that discursive frameworks have considerable power in shaping and directing policy and strategy. In such a reading, then, one must concede, as many constructivists do, that discourse has a constitutive power: '[p]olicy statements should be "treated as actions...rather than policies," because policies are instruments that result in action' (Onuf 2001: 79). This has been made clear throughout this work: with Blair in 2001 and 2002 and his placing of a framework of liberal peace theory onto the future course of the mission, with Reid and Browne persisting in framing the mission in Helmand as one of 'reconstruction' despite all evidence to the contrary, with Hutton reconfiguring the rationale for the conflict from one of global security and for the sake of Afghanistan to one primarily concerned with national security and counter-terrorism, and with the Cameron Government's reframing of Afghanistan within the context of a narrow conception of the national interest.

In each case, the narrative pathway taken had ramifications that narrowed the rhetorical options available and contributed to the gradual altering of Britain's strategic posture in Afghanistan. It is also important to recognise the limits of communication practices in strategic affairs, however; as Stuart Griffin (2011: 331) has argued, '[d]octrine doesn't shape policy and strategy: policy and strategy shape doctrine'. Treating policy statements as actions will not change the fact that words cannot substitute for action, just as rhetoric cannot meaningfully overcome a lack of autonomy in policymaking. Of course, the conclusion this work has arrived at is that the United Kingdom's policy narratives for Afghanistan, directed by transnational processes and in response to the transnational dilemma, have sought to do precisely that. Communication on Afghanistan has effectively sidestepped the problems of British strategy and, in the process, has assumed a mantle of being 'strategic' in the absence of an easily defined continuum of interests, policies and strategies. The essential paradox here is that because strategic communication arose in spite of, and not in support of such a continuum, it cannot but obscure – and at worst further compound – the strategic shortcomings of the British state. This is how it should be analysed – empirically and then theoretically. Whilst interests do abide eternally (and therefore should be central to strategic studies), their substance and location, and our understanding of what they consist of and where they reside, remains uncertain as they are malleable to the processes of transnationalisation through which they are located and re-located. There is very little that is conceptually clear about 'national interests' under such circumstances (beyond its rhetorical ambiguity and therefore its political utility), and it is this confusion that gives rise to ideas of transnational dilemmas and, I have argued, has precipitated the need for strategic communication. Explaining the shortcomings of British strategy by reference to this dynamic has exposed this gap in strategic studies and the limitations of SC. As long as war is considered a rational pursuit, there must be a clear conceptual link between policy and strategy. If the policy is unstatable, however, the possibility of strategies becoming decoupled from it is highly likely. If an unstatable policy is also fundamental and unavoidable, the potential for improving strategic thinking is diminished, not least since the protection of an unstated yet essential policy can (and apparently has) become a strategic aim in itself. This is a task for which strategic communication is well-suited, but also one that, by design, further obscures the strategic quandaries at the heart of British defence policy.

Bibliography

Books and Journal Articles

Alexander, J. 2011. *Performance and Power.* Cambridge: Polity.

Barkawi, T. 2006. 'Terrorism and North-South Relations'. *The RUSI Journal* 151(1): 54–8.

Baumann, A. 2009. 'Constructive Friction or Petty Turf Wars? Organisational Resistance to the Integration of Defence, Diplomacy and Development'. *Chatham House.* http://www.chathamhouse.org/sites/files/chathamhouse/public/Research/International%20Security/1109esdf_baumann.pdf.

Bearden, M. 2001. 'Afghanistan, Graveyard of Empires'. *Foreign Affairs* 80(6): 17–30.

Beech, M. 2006. *The Political Philosophy of New Labour.* London: Tauris.

Belcher, O. 2012. 'The Best-Laid Schemes: Postcolonialism, Military Social Science, and the Making of US Counterinsurgency Doctrine, 1947–2009'. *Antipode* 44(1): 258–63.

Betz, D. 2011. 'Communication Breakdown: Strategic Communications and Defeat in Afghanistan'. *Orbis* 55(4): 613–30.

Betz, D. and Cormack, A. 2009. 'Iraq, Afghanistan and British Strategy'. *Orbis* 53(2): 319–36.

Biggar, N. 2011. 'The Invasion of Iraq: What are the Morals of the Story?'. *International Affairs* 87(1): 29–37.

Bird, T. and Marshall, A. 2011. *Afghanistan: How the West Lost its Way.* London: Yale University Press.

Birkle, G., O'Hanlon, M. and Sherjan, H. 2011. 'Toward a Political Strategy for Afghanistan'. *The Brookings Institution.*

Blackwill, R.D. 2011. 'Plan B in Afghanistan: Why a De Facto Partition is the Least Bad Option'. *International Affairs* 90(1): 42–50.

Blair, T. 2010. *A Journey.* London: Hutchinson.

Bower, T. 2007. *Gordon Brown: Prime Minister.* London: Harper Perennial.

Brown, R. 2003.'Spinning the War: Political Communications, Information Operations and Public Diplomacy in the War on Terrorism', in *War and the Media*, edited by D.K. Thussu and D. Freedman. London: Sage, pp.87–100.

Buzan, B. 2004. *From International to World Society?: English School Theory and the Social Structure of Globalisation.* Cambridge: Cambridge University Press.

Buzan, B., Wæver, O. and De Wilde, J. 1998. *Security: A New Framework for Analysis.* Boulder, CO: Lynne Rienner.

Buzan, B. and Wæver, O. 2003. *Regions and Powers: The Structure of International Security*. Cambridge: Cambridge University Press.

Cassidy, R.M. 2010. 'The Afghanistan Choice'. *The RUSI Journal* 155(4): 38–44.

Cavanagh, M. 2012. 'Ministerial Decision-Making in the Run-Up to the Helmand Deployment'. *The RUSI Journal* 157(2): 48–54.

Chaudhuri, R. and Farrell, T. 2011. 'Campaign Disconnect: Operational Progress and Strategic Obstacles in Afghanistan, 2009–2011'. *International Affairs* 87(2): 271–96.

Chin, W. 2009. 'The United Kingdom and the War on Terror: The Breakdown of National and Military Strategy'. *Contemporary Security Policy* 30(1): 125–46.

Clarke, M. 2007. 'Foreign Policy', in *Blair's Britain, 1997–2007,* edited by A. Seldon. Cambridge: Cambridge University Press, pp.593–614.

Clarke, M. 2012. 'The Helmand Decision', in *The Afghan Papers: Committing Britain to War in Helmand, 2005–06*, edited by M. Clarke. Abingdon: RUSI.

Clausewitz, C. 1982. 'Anatol Rapoport', in *On War*, translated by J. Graham. London: Penguin.

Cornish, P. 2004. 'NATO: The Practice and Politics of Transformation'. *International Affairs* 80(1): 63–74.

Cornish, P. and Dorman, A.M. 2009a. 'Blair's Wars and Brown's Budgets: From Strategic Defence Review to Strategic Decay in Less than a Decade'. *International Affairs* 85(2): 247–61.

Cornish, P. and Dorman, A.M. 2009b. 'National Defence in the Age of Austerity'. *International Affairs* 85(4): 733–53.

Cornish, P. and Dorman, A.M. 2010. 'Breaking the Mould: The United Kingdom Strategic Defence Review 2010'. *International Affairs* 86(2): 395–410.

Cornish, P. and Dorman, A.M. 2011. 'Dr Fox and the Philosopher's Stone: The Alchemy of national Defence in the Age of Austerity'. *International Affairs* 87(2): 335–53.

Cornish, P. and Dorman, A.M. 2012. 'Smart Muddling Through: Rethinking UK National Strategy Beyond Afghanistan'. *International Affairs* 88(2): 213–22.

Cornish, P., Lindley-French, J. and Yorke, C. 2011. *Strategic Communications and National Strategy*. London: Chatham House.

Cowper-Coles, S. 2011. *Cables from Kabul: The Inside Story of the West's Afghanistan Campaign.* London: Harper.

Dalacoura, K. 2012. 'The 2011 uprisings in the Arab Middle East: Political Change and Geopolitical Implications'. *International Affairs* 88(2): 63–79.

Dannatt, R. 2011. *Leading from the Front.* London: Corgi.

Desch, M.C. 2007/08. 'America's Liberal Illiberalism: The Ideological Origins of Overreaction in U.S. Foreign Policy'. *International Security* 32(3): 7–43.

Dimitriu, G.R. 2012. 'Winning the Story War: Strategic Communication and the Conflict in Afghanistan'. *Public Relations Review.* 38(2): 195–207.

Dodge, T. 2010. 'The Ideological Roots of Failure: The Application of Kinetic Neo-liberalism to Iraq'. *International Affairs* 86(6): 1269–86.

Dodge, T. 2011. 'Domestic Politics and State-building', in *Afghanistan: To 2015 and Beyond*, edited by T. Dodge and N. Redman. London: Routledge.

Dorman, A. 2012. 'NATO's 2012 Chicago Summit: A Chance to Ignore the Issues Once Again?'. *International Affairs* 88(2): 301–12.

Doyle, M. 1986. 'Liberalism and World Politics'. *The American Political Science Review* 80(4): 1151–69.

Duffield, M. 2001. *Global Governance and the New Wars: The Merging of Development and Security.* London: Zed.

Dunn, D.H. 2003. 'Myths, Motivations and "Misunderestimations": The Bush Administration and Iraq'. *International Affairs* 79(2): 279–97.

Dunn, D.H. 2008. 'The Double Interregnum: UK–US relations Beyond Blair and Bush'. *International Affairs* 84(6): 1131–43.

Edmunds, T. 2010. 'The Defence Dilemma in Britain'. *International Affairs* 86(2): 377–94.

Edmunds, T. 2012. 'British Civil-military Relations and the Problem of Risk'. *International Affairs* 88(2): 265–82.

Edmunds, T. 2014. 'Complexity, Strategy and the National Interest'. *International Affairs* 90(3): 525–39.

Edmunds, T., Gaskarth, J. and Porter, R. 2014. 'Introduction: British Foreign Policy and the National Interest'. *International Affairs* 90(3): 503–7.

Eriksson, J. and Rhinard, M. 2009. 'The Internal-External Security Nexus: Notes on an Emerging Research Agenda'. *Cooperation and Conflict* 44(3): 243–67.

Etzioni, A. 2012. 'The Case for Decoupled Armed Interventions'. *Global Policy* 3(1): 85–93.

Farrell, T. and Rynning, S. 2010. 'NATO's Transformation Gaps: Transatlantic Differences and the War in Afghanistan'. *The Journal of Strategic Studies* 33(5): 673–99.

Farrell, T. and Gordon, S. 2009. 'COIN Machine: The British Military in Afghanistan'. *The RUSI Journal* 154(3): 18–25.

Felbab-Brown, V. 2009. 'Peacekeepers Among Poppies: Afghanistan, Illicit Economies and Intervention'. *International Peacekeeping* 16(1): 100–14.

Fergusson, J. 2008. *A Million Bullets: The Real Story of the British Army in Afghanistan.* London: Bantam.

Freedman, L. 2006a. *The Transformation of Strategic Affairs*. Abingdon: Routledge.

Freedman, L. 2006b. 'The Special Relationship, Then and Now'. *Foreign Affairs* 85(3): 61–73.

Freedman, L. 2007. 'Defence', in *Blair's Britain, 1997–2007*, edited by A. Seldon. Cambridge: Cambridge University Press, pp.615–32.

Freedman, L. 2013. *Strategy: A History.* Oxford: Oxford University Press.

Gamble, A. 2003. *Between Europe and America: The Future of British Politics.* London: Palgrave.

Gaskarth, J. 2014. 'Strategizing Britain's Role in the World'. *International Affairs* 90(3): 559–81.

Gentile, G. 2011a. 'COIN is Dead: U.S. Army Must Put Strategy Over Tactics'. *Small Wars Journal.* http://smallwarsjournal.com/blog/coin-is-dead-us-army-must-put-strategy-over-tactics.

Gentile, G. 2011b. 'Beneficial War: The Conceit of American Counterinsurgency'. *Small Wars Journal.* http://smallwarsjournal.com/blog/beneficial-war-the-conceit-of-american-counterinsurgency.

Gilboa, E. 2008. 'Searching for a Theory of Public Diplomacy'. *Annals of the American Academy of Political and Social Science* 616(1): 55–77.

Gilmore, J. 2014. 'The Uncertain Merger of Values and Interests in UK Foreign Policy'. *International Affairs* 90(3): 541–57.

Gowing, N. 2009. *'Skyful of Lies' and Black Swans: The New Tyranny of Shifting Information Power in Crises.* Oxford: Reuters Institute for the Study of Journalism.

Gray, C.S. 2007. *War, Peace and International Relations: An Introduction to Strategic History.* London: Routledge.

Gray, C.S. 2008. 'Britain's National Security: Compulsion and Discretion'. *The RUSI Journal* 153(6): 13–18.

Gray, C.S. 2010. 'Strategic Thoughts for Defence Planners'. *Survival* 52(3): 159–78.

Griffin, S. 2011. 'Iraq, Afghanistan and the Future of British Military Doctrine: From Counterinsurgency to Stabilization'. *International Affairs* 87(2): 317–33.

Hain, P. 2001. *The End of Foreign Policy? Britain's Interests, Global Linkages and Natural Limits.* Fabian Society, Green Alliance and Royal Institute of International Affairs, London.

Harvey, C. and Wilkinson, M. 2009. 'The Value of Doctrine'. *The RUSI Journal* 154(6): 26–31.

Hastings, M. 2013. *The Operators: The Wild and Terrifying Inside Story of America's War in Afghanistan.* London: Phoenix.

Helmus, T.C., Paul, C. and Glenn, R.W. 2007. *Enlisting Madison Avenue: The Marketing Approach to Earning Popular Support in Theaters of Operation.* Santa Monica: Rand.

Hill, C. 2005. 'Putting the World to Rights: Tony Blair's Foreign Policy Mission', in *The Blair Effect 2001–5*, edited by A. Seldon and D. Kavanagh. Cambridge: Cambridge University Press.

Holland, J. 2008. 'The Way Ahead in Afghanistan'. *The RUSI Journal* 153(3): 46–50.

Howell, J. 2010. 'National Security Concerns Continue to Dictate Britain's Government Aid and Development Agendas'. Blogs.LSE.ac.uk http://eprints.lse.ac.uk/39736/1/blogs.lse.ac.uk-National_security_concerns_continue_to_dictate_Britains_government_aid_and_development_agendas.pdf.

Hudson, K. 1978. *The Language of Modern Politics.* London: Macmillan.

Hudson, L., Owens, C.S. and Flannes, M. 2011. 'Drone Warfare: Blowback from the New American Way of War'. *Middle East Policy* 18(3): 122–32.

Hunt, N. and Stevens, A. 2004. 'Whose Harm? Drug Reduction and the Shift to Coercion in UK Drug Policy'. *Social Policy and Society* 3(4): 333–42.

Jackson, P. 2010. 'SSR and Post-Conflict Reconstruction: The Armed Wing of State Building?', in *The Future of Security Sector Reform*, edited by M. Sedra. CIGI: Ontario, pp.118–35.

Janis, I.L. 1972. *Groupthink: Psychological Studies of Policy Decisions and Fiascoes*. Boston, MA: Houghton Mifflin.

Johnson, R. 2012. 'British Approaches to Pacification in Afghanistan, 1842–1880'. *Peace Research Institute Oslo (PRIO)*. www.prio.org/utility/Download. ashx?x=221.

Jones, D.M. and Smith, M.L.R. 2010. 'Whose Hearts and Whose Minds? The Curious Case of Global Counter-Insurgency'. *Journal of Strategic Studies* 33(1): 81–121.

Jones, S.G. 2008. 'The Rise of Afghanistan's Insurgency: State Failure and Jihad'. *International Security* 32(4): 7–40.

Kant, I. 2002. *Groundwork for the Metaphysics of Morals*, edited by Allen W. Wood. London: Yale University Press.

King, A. 2010a. 'The Power of Politics: Hamkari and the Future of the Afghan War'. *The RUSI Journal* 155(6): 68–74.

King, A. 2010b. 'Understanding the Helmand Campaign: British Military Operations in Afghanistan'. *International Affairs* 86(2): 311–32.

King, A. 2011. *The Transformation of Europe's Armed Forces: From the Rhine to Afghanistan*. Cambridge: Cambridge University Press.

King, A. 2011. 'Military Command in the Last Decade'. *International Affairs* 87(2): 377–96.

Kreps, S. 2010. 'Elite Consensus as a Determinant of Alliance Cohesion: Why Public Opinion Hardly Matters for NATO-led Operations in Afghanistan'. *Foreign Policy Analysis* 6(3): 191–215.

Kriner, D.L. and Wilson, G. 2010. 'Elites, Events and British Support for the War in Afghanistan'. *APSA Annual Meeting Paper.* http://papers.ssrn.com/sol3/papers.cfm?abstract_id=1644301.

Layton, P. 2010. 'The Idea of Grand Strategy'. *RUSI Journal,* August-September 2012, 157(4): 56–61.

Ledwidge, F. 2011. *Losing Small Wars: British Military Failure in Iraq and Afghanistan*. London: Yale University Press.

Ledwidge, F. 2013. *Investment in Blood: The True Cost of Britain's Afghan War.* London: Yale University Press.

Lindsay, J.M. 2011. 'George W Bush, Barack Obama and the Future of US Global Leadership'. *International Affairs* 87(4): 765–79.

Mackay, A. and Tatham, S. 2009. 'Behavioural Conflict: From General to Strategic Corporal: Complexity, Adaptation and Influence'. *The Shrivenham Papers.* Defence Academy of the United Kingdom, Shrivenham.

Marsden, P. 2003. 'Afghanistan: The Reconstruction Process'. *International Affairs* 79(1): 91–105.

Mccrisken, T. 2011. 'Ten Years On: Obama's War on Terrorism in Rhetoric and Practice'. *International Affairs* 87(4): 781–801.

Miller, P.D. 2011. 'Finish the Job: How the War in Afghanistan Can Be Won'. *Foreign Affairs* 90(1): 51–65.

Monten, J. 2005. 'The Roots of the Bush Doctrine: Power, Nationalism, and Democracy Promotion in U.S. Strategy'. *International Security* 29(4): 112–56.

Newton, P., Colley, P. and Sharpe, A. 2010. 'Reclaiming the Art of British Strategic Thinking'. *RUSI Journal*, February–March, 155(1): 44–51.

Nissen, T.E. 2012. 'Narrative Led Operations: Put the Narrative First'. *Small Wars Journal.* http://smallwarsjournal.com/jrnl/art/narrative-led-operations-put-the-narrative-first.

Onuf, N. 2001. 'Speaking of Policy', in *Foreign Policy in a Constructed World,* edited by V. Kubalkova. London: M.E. Sharpe, pp.77–95.

O'Neal, J.R., O'Neal, F.H., Maoz, Z. and Russett, B. 1996. 'The Liberal Peace: Interdependence, Democracy, and International Conflict, 1950–1985'. *Journal of Peace Research* 33(1): 11–28.

Paris, R. 2009. *At War's End: Building Peace After Civil Conflict.* Cambridge: Cambridge University Press.

Paris, R. 2010. 'Saving Liberal Peacebuilding'. *Review of International Studies* 36(2): 337–65.

Patterson, E. 2012. 'Obama and Sustainable Democracy Promotion'. *International Studies Perspectives* 13(1): 26–42.

Paul, C. 2011. *Strategic Communication.* Santa Barbara, CA: Praeger.

Porter, P. 2009. *Military Orientalism: Eastern War through Western Eyes.* London: Hurst.

Porter, P. 2010a. 'Last Charge of the Knights? Iraq, Afghanistan and the Special Relationship'. *International Affairs* 86(2): 355–75.

Porter, P. 2010b. 'Why Britain Doesn't Do Grand Strategy'. *The RUSI Journal* 155(4): 6–12.

Porter, R. 2014. 'Why America?', in *British Foreign Policy and the National Interest: Identity, Strategy and Security,* edited by T. Edmunds, J. Gaskarth and R. Porter. London: Palgrave.

Price, M.E. 2009. 'End of Television and Foreign Policy'. *Annals of the American Academy of Political and Social Science* 625(1): 196–204.

Rawls, J. 1999. *The Law of Peoples with "The Idea of Public Reason Revisited"* London: Harvard University Press.

Ricks, T.E. 2009. *The Gamble: General Petraeus and the Untold Story of the American Surge in Iraq, 2006–2008.* London: Allen Lane.

Ringsmose, J. and Borgesen, B.K. 2011. 'Shaping Public Attitudes towards the Deployment of Military Power: NATO, Afghanistan and the Use of Strategic Narratives'. *European Security* 20(4): 505–28.

Roenfeldt, C. 2011. 'Productive War: A Re-Conceptualisation of War'. *The Journal of Strategic Studies* 34(1): 39–62.

Rynning, S. 2012. *NATO in Afghanistan: The Liberal Disconnect.* Stanford, CA: Stanford University Press.

Sanger, M. 2012. *Confront and Conceal: Obama's Secret Wars and Surprising Use of American Power.* New York: Crown.

Seldon, A. 2005. *Blair.* London: Free Press.

Seldon, A. and Lodge, G. 2011. *Brown at 10.* London: Biteback.

Short, C. 2010. 'Foreword', in *The Future of Security Sector Reform*, edited by M. Sedra. CIGI: Ontario, pp.10–14.

Simon, L. and Rogers, J. 2011. 'British Geostrategy for a New European Age'. *RUSI Journal* 156(2): 52–8.

Simpson, E. 2012. *War From the Ground Up: Twenty-First Century Combat as Politics.* London: C. Hurst.

Smith, P. 2005. *Why War?: The Cultural Logic of Iraq, the Gulf War, and Suez.* Chicago, IL: University of Chicago Press.

Smith, P. and Riley, A. 2009. *Cultural Theory: An Introduction.* Oxford: Blackwell.

Smith, R. 2006. *The Utility of Force: The Art of War in the Modern World.* London: Penguin.

Snyder, J.T. 2011. 'Counterinsurgency Vocabulary and Strategic Success'. *Military Review* 91(6): 23–8.

Snyder, R.C., Bruck, H.W. and Sapin, B. 2002. 'Decision-Making as an Approach to the Study of International Politics', in Snyder, R.C., Bruck, H.W. and Sapin, B., *Foreign Policy Decision-Making.* Basingstoke: Palgrave Macmillan, pp.21–152.

Stimson, G. 2000. '"Blair Declares War'. or the Unhealthy State of British Drugs Policy'. *UK Harm Reduction Alliance.* http://www.ukhra.org/stimsonspeech.html.

Strachan, H. 2005. 'The Lost Meaning of Strategy'. *Survival* 47(3): 33–54.

Strachan, H. 2006. 'Making Strategy: Civil-military relations after Iraq'. *Survival* 48(3): 59–82.

Strachan, H. 2008. 'Strategy and the Limitation of War'. *Survival* 50(1): 31–54.

Suhrke, A. 2011. *When More is Less: The International Project in Afghanistan.* London: Hurst.

Suhrke, A. 2013. 'Statebuilding in Afghanistan: A Contradictory Engagement'. *Central Asian Survey* 32(3): 271–86.

Tatham, S. 2008. *Strategic Communication: A Primer.* Defence Academy of the United Kingdom, Shrivenham.

West, B. 2011. 'The Way Out of Afghanistan'. *Military Review* 91(2): 89–95.

Williams, M.J. 2011. 'Empire Lite Revisited: NATO, the Comprehensive Approach and State-building in Afghanistan'. *International Peacekeeping* 18(1): 64–78.

Williams, M.J. 2009. *NATO, Security and Risk Management: From Kosovo to Kandahar.* London: Routledge.

Williams, P. 2004. 'Who's Making UK Foreign Policy?'. *International Affairs* 80(5): 909–29.

Williams, P.D. 2005. *British Foreign Policy under New Labour, 1997–2005.* Basingstoke: Palgrave MacMillan.

Williams, S. 2012. 'The Role of the National Interest in the National Security Debate'. *Royal College of Defence Studies.*

Wolfers, A. 1962. *Discord and Collaboration: Essays on International Politics.* Baltimore, MD: Johns Hopkins.

Wright, J. 2009. 'Out in Force: Military Operations in Helmand Province'. *Jane's Intelligence Review* 21(10): 20–29.

Media Articles

Astill, J. 2003 (21 November). 'Plea for security rethink as French aid worker is buried: UN says relief work in Afghanistan cannot continue on existing terms'. *The Guardian*, p.20.

Baker, G. 2005 (29 December). 'NATO facing a critical test of its resolve from resurgent Taleban'. *The Times*, p.36.

Batty, D. 2010 (22 May). 'Liam Fox calls for faster UK withdrawal from Afghanistan'. *The Guardian.* http://www.theguardian.com/politics/2010/may/22/liam-fox-troop-withdrawal-afghanistan.

BBC. 2002 (12 December). 'UK targets Afghan heroin production'. *BBC.* http://news.bbc.co.uk/1/hi/uk/2568027.stm.

BBC. 2003 (1 May). 'Afghanistan 'moves to stability''. *BBC.* http://news.bbc.co.uk/1/hi/world/south_asia/2991121.stm.

BBC. 2004 (7 December). 'Karzai warns of Afghan dangers'. *BBC.* http://news.bbc.co.uk/1/hi/world/south_asia/4076331.stm.

BBC. 2006 (24 April). 'UK troops 'to target terrorists''. *BBC.* http://news.bbc.co.uk/1/hi/uk/4935532.stm.

BBC. 2007 (28 January). 'Blair: "I'm going to finish what I started". *The Politics Show.* http://news.bbc.co.uk/1/hi/programmes/politics_show/6293605.stm.

BBC. 2008 (10 October). 'NATO to attack Afghan drugs labs'. *BBC.* http://news.bbc.co.uk/1/hi/7663204.stm.

BBC. 2009 (21 March). 'Envoy damns US drugs effort'. *BBC.* http://news.bbc.co.uk/1/hi/world/south_asia/7957237.stm.

BBC. 2013 (16 December). 'Afghanistan mission accomplished, says David Cameron'. *BBC.* http://www.bbc.co.uk/news/uk-politics-25398608.

Beeston, R. 2002 (30 November). 'British troops to join US in wider Afghanistan role'. *The Times*, p.20.

Beeston, R. 2004a (30 March). 'More British troops for Afghanistan'. *The Times*, p.14.

Beeston, R. 2004b (1 April). 'Afghanistan "needs £15bn to rebuild state"'. *The Times*, p.20.

Bentham, M. 2005 (29 May). 'British to assault stronghold: Commander warns that peace-enforcing task in Afghanistan could occupy troops "for a generation"'. *The Observer*, p.21.

Black, I. and White, M. 2004 (29 June). 'NATO pledge for Afghanistan'. *The Guardian*, p.11.

Brown, C. 2004 (26 May). 'Iraq Crisis: Afghanistan is not forgotten, says PM'. *The Independent*, p.6.

Burke, J. 2002 (28 April). 'Troops fight boredom in the war on terror'. *The Observer*, p.18.

Burke, J. 2003 (16 November). 'Stronger and more deadly, the terror of the Taliban is back'. *The Observer*, p.23.

Burke, J. 2004 (5 December). 'British troops wage war on Afghan drugs'. *The Guardian*. http://www.theguardian.com/world/2004/dec/05/afghanistan. observerpolitics.

Burleigh, M. 2009 (20 February). 'Do our men die in Afghanistan to show willing in Washington?'. *The Daily Telegraph*, p.22.

Carr, S. 2002 (19 July). 'Mission Accomplished – Last British combat troops leave Afghanistan'. *The Independent*, p.8.

Castle, S. 2006 (17 January). 'War in Afghanistan: Why an expansion of NATO's role has divided the Dutch'. *The Independent*, p.2.

Cavanagh, M. 2010 (17 November). 'Inside the Anglo-Saxon war machine'. *Prospect*. http://www.prospectmagazine.co.uk/magazine/inside-the-anglo-saxon-war-machine.

Coghlan, T. 2005a (31 August). 'Afghan opium farmers' anger as West threatens crop controls'. *The Telegraph*. http://www.telegraph.co.uk/news/worldnews/asia/afghanistan/1497313/Afghan-opium-farmers-anger-at-West-threatens-crop-controls.html.

Coghlan, T. 2005b (15 December). 'Challenge for British troops as Afghans sow a new opium crop'. *The Telegraph*. http://www.telegraph.co.uk/news/worldnews/asia/afghanistan/1505553/Challenge-for-British-troops-as-Afghans-sow-a-new-opium-crop.html.

Coughlin, C. 2008a (26 June). 'Why the top brass are breaking ranks'. *The Daily Telegraph*, p.22.

Coughlin, C. 2008b (30 May). 'War is hell, it is never going to be a politically correct pastime Britain's decision to ban the use of cluster bombs will only endanger our troops'. *The Daily Telegraph*, p.25.

Coughlin, C. 2008c (31 October). 'Now more than ever, Britain needs a plan for Afghanistan: John Hutton, the new Defence Secretary, is right to concentrate on defeating the Taliban insurgency'. *The Daily Telegraph*, p.26.

Coughlin, C. 2009a (1 May). 'Brown's cost-cutting will put our soldiers' lives at further risk'. *The Daily Telegraph*, p.30.

Coughlin, C. 2009b (2 July). 'The American surge in Afghanistan is an embarrassment to Britain'. *The Daily Telegraph*. blog.

Cook, R. 2004 (2 July). 'We owe it to the people of Afghanistan to repair their broken, blighted land'. *The Independent*, p.35.

Ctv.ca staff. 2007 (12 May). 'Canada stays out of Afghan opium poppy harvest'. *CTV.*   http://www.ctvnews.ca/canada-stays-out-of-afghan-opium-poppy-harvest-1.241002.

Daily Telegraph. 2006 (19 January). 'NATO must not weaken on Afghan terror front'. *The Daily Telegraph*, p.23.

Daily Telegraph. 2011 (4 February). '(SBU) HMG considers role in Afghanistan after August presidential election – Debate continues on helicopter And troop levels'. *The Daily Telegraph.* http://telegraph.co.uk/news/wikileaks-files/london-wikileaks/8305296/sbu-hmg-considers-role-in-afghanistan-after-august-presidential-election-debate-continues-on-helicopter-and-troop-levels.html.

Dawi, A. 2014 (6 March). 'Despite Massive Taliban Death Toll No Drop in Insurgency'. *Voice of America* http://www.voanews.com/content/despite-massive-taliban-death-toll-no-drop-in-insurgency/1866009.html.

Dejevsky, M. 2001 (1 December). 'Campaign against terrorism: Hopes of peace deal rise as talks hang in balance'. *The Independent*, p.8.

Eagar, C. 2006 (6 August). 'The Great British Opium Swindle'. *The Mail on Sunday*, p.32.

Elliott, F. 2001 (18 November). 'Britain and US plan to stop heroin trade by buying Afghan opium crop'. *The Telegraph.* http://www.telegraph.co.uk/news/worldnews/asia/afghanistan/1362740/Britain-and-US-plan-to-stop-heroin-trade-by-buying-Afghan-opium-crop.html.

Evans, M. 2008a (6 June). 'General given short shrift after complaint that soldiers earn less than traffic wardens'. *The Times*, p.22.

Evans, M. 2008b (10 July). 'Fighting two wars takes Army close to breaking point'. *The Times*, p.21.

Evans, M. 2009 (30 March). 'Expanded mission in Afghanistan exposed our Basra forces to attack'. *The Times*, p.30.

Evans, M. and Coates, S. 2009 (27 March). 'Army is poised for Afghanistan surge; Britain set to follow Obama and send more troops'. *The Times*, p.1.

Evans, M and Bone, J. 2001 (20 December). 'Britain sets time limit on peacekeeping'. *The Times.*

Fickling, D. 2006 (8 August). 'US defends opium policy despite violence'. *The Guardian.*   http://www.theguardian.com/world/2006/aug/08/afghanistan.politics.

Fisk, R. 2003 (5 February). 'Don't mention the war in Afghanistan'. *The Independent*, p.17.

Fox, R. 2008 (11 December). 'Hutton slashes defence spending across forces'. *London Evening Standard*, p.18.

Fox, L. 2008 (6 April). 'If Brown undermines NATO, he will have a fight on his hands'. *The Sunday Telegraph*, p.25.

Fox, N. 2004 (29 March). 'Afghanistan puts elections on hold'. *The Independent*, p.25.

Gall, C. 2003 (23 April). 'Taleban revival haunts 'forgotten' Afghanistan'. *The Times*, p.13.

Gilligan, A. 2002 (19 May). 'Britain's armed forces are spinning above their weight'. *The Sunday Telegraph*, p.15.

Goldenberg, S. 2002 (21 November). 'US must put Afghanistan back together: Focus shifts from security and combat to reconstruction'. *The Guardian*, p.15.

Graham-Harrison, E. 2012 (20 May). 'Taliban destroy poppy fields in surprise clampdown on Afghan opium growers'. *The Guardian.* http://www. theguardian.com/world/2012/may/20/taliban-destroy-poppy-afghan-opium.

Grice, E. 2009 (21 August). 'A Christian soldier's fight for truth'. *The Daily Telegraph*, p.17.

Guardian. 2006 (25 January). 'Reid to announce new British troops for Afghanistan'. *The Guardian.*

Guardian. 2007 (13 August). 'We are making a difference in Afghanistan, insists Browne'. *The Guardian.*

Guardian. 2009a (15 January). 'The end of NATO?'. *The Guardian.*

Guardian. 2009b (19 February). 'US urges NATO allies must shoulder more responsibility in Afghanistan'. *The Guardian.*

Hames, T. 2007 (17 December). 'Iraq – the biggest, and best, story of the year'. *The Times*, p.19.

Harding, L. 2002 (27 February). 'A new war is brewing in Afghanistan: Unless British troops stay for the long haul, fighting may reignite'. *The Guardian*, p.20.

Harding, L. 2003 (9 June). 'US helicopters in secret mission to spray Afghanistan's opium fields'. *The Guardian.* http://www.theguardian.com/world/2003/jun/09/afghanistan.usa.

Harding, T. 2005 (14 June). 'More British troops for Afghanistan'. *The Daily Telegraph*, p.11.

Helm, T. 2006 (28 November). 'Blair urges NATO allies to do more fighting in Afghanistan'. *The Daily Telegraph*, p.17.

Hemming, J. 2007 (8 October). 'U.S. and Afghan officials meet on drug spray chemical'. *Reuters.* http://uk.reuters.com/article/2007/10/08/uk-afghan-drugs-idUKISL10021720071008.

Hennessey, P. 2010 (26 June). 'Prime Minister David Cameron adds pressure on President Hamid Karzai at G8 in Canada'. *The Telegraph.* http://www. telegraph.co.uk/news/worldnews/northamerica/canada/7856522/Prime-Minister-David-Cameron-adds-pressure-on-President-Hamid-Karzai-at-G8-in-Canada.html.

Hinsliff, G. 2009 (19 July). 'Bitter fallout as Brown and the generals caught in war games'. *The Observer*, p.18.

Hollander, G. 2010 (5 September). 'Labour's policy on Iraq was 'fatally flawed'. says former Army chief'. *The Observer.* http://www.theguardian.com/politics/2010/sep/05/richard-dannatt-defence-spending.

Hopkins, N. 2001 (4 October). 'All seized UK heroin traced to Afghans'. *The Guardian.*   http://www.theguardian.com/uk/2001/oct/04/drugsandalcohol. afghanistan.

Howells, K. 2007 (24 August). 'Our intervention in Afghanistan has nothing to do with jingoism'. *The Guardian*, p.43.

Hussain, Z. 2003 (20 October). 'Taleban return from shadows to boast of victories'. *The Times*, p.15.

Independent. 2002 (13 April). 'We should be peace-keeping not fighting in Afghanistan'. *The Independent*, p.3.

Independent. 2006 (17 January). 'Moral duty and self-interest'. *The Independent*, p.26.

Independent. 2007 (14 August). 'Politicians must accept the reality on the ground'. *The Independent*, p.32.

Jenkins, S. 2006 (1 February). 'Blair's latest expedition is a Lawrence of Arabia fantasy'. *The Guardian*, p.31.

Jones, A. 2006 (17 December) 'Drug war, Taliban, poppies are all in full flower / Opium, thugs bloom under U.S. policies in Afghanistan war'. *SFGate.* http:// www.sfgate.com/opinion/article/Drug-war-Taliban-poppies-are-all-in-full-flower-2465331.php.

Kirkup, J. 2008a (4 April). 'Brown may send more British troops to Afghanistan'. *The Daily Telegraph*, p.20.

Kirkup, J. 2008b (6 June). 'Front-line troops paid less than traffic wardens, says Army chief'. *The Daily Telegraph*, p.6.

Kirkup, J. 2009 (19 February). 'Explaining Britain's Afghanistan mission proves tricky'. *The Daily Telegraph.* Blog.

Kirkup, J. 2010 (8 July). 'Events, dear boy: David Cameron's Afghan 'timetable'. might slip'. *The Daily Telegraph.* Blog.

Kirkup, J. and Spillius, A. 2011 (23 June). 'British pull out of Afghanistan to speed up'. *The Daily Telegraph*, p.1,2.

La Guardia, A. 2003 (7 October). 'NATO moves to prop up Afghan leader'. *The Daily Telegraph*, p.14.

Lamb, C. 2005 (24 April). 'British troops to target Afghan opium trade'. *The Times*, p.2.

Loof, S. 2003 (29 October). 'U.N. Agency Warns Afghanistan Over Opium'. *The Guardian/RAWA.* http://www.rawa.org/opium4.htm.

Loyd, A. 2004 (7 August). 'Afghan poppy farmers give up their daughters to pay off opium debts'. *The Times*, p.20.

Loyd, A. and Khel, G. 2003 (29 August). 'Poppy crop flourishes despite anti-drug drive'. *The Times*, p.17.

Macintyre, D. 2002 (11 June). 'It is in our own selfish interest to rebuild a stable society in Afghanistan'. *The Independent*, p.16.

Macshane, D. 2009 (21 August). 'We can't abandon Afghanistan'. *The Guardian.*

Maddox, B. 2004 (28 June). 'Future of NATO hangs on its help in Afghanistan'. *The Times*, p.12.

Malloch-Brown, M. 2007 (29 November). 'Letters – Taliban no longer a credible threat'. *The Independent.* http://www.independent.co.uk/voices/letters/letters-political-funding-760853.html.

Massoud, A.Z. 2007 (2 September). 'Leave it to us to end the poppy curse'. *The Sunday Telegraph*, p.22.

Mcgrory, D. and Hussein, Z. 2006 (6 April). 'MPs fear for British forces in battle with opium gangs'. *The Times*, p.31.

Meo, N. 2003 (2 August). 'Blair drug fiasco'. *The Daily Mirror*, p.10.

Meo, N. 2005 (15 May). 'Bush's wars: And in Afghanistan, the Taliban rises again for fighting season'. *The Independent on Sunday*, p.19.

Nelson, F. 2013 (07 November). 'We must fund the Armed Forces properly – before disaster strikes'. *The Telegraph.* http://www.telegraph.co.uk/news/uknews/defence/10433063/We-must-fund-the-Armed-Forces-properly-before-disaster-strikes.html.

Norton-Taylor, R. 2002 (19 March). '1,700 UK troops for Afghanistan: 1,700 UK troops to fight in Afghan war'. *The Guardian*, p.1.

Norton-Taylor, R. 2005a (6 July). 'UK seeks to free troops for Afghanistan'. *The Guardian*, p.15.

Norton-Taylor, R. 2005b (4 November). 'Britain isolated over role in Afghanistan: Allies reluctant to get involved in war on terror: Tribal feuds and opium trade hinder peacekeeping'. *The Guardian*, p.21.

Norton-Taylor, R. 2006 (27 January). 'Britain to commit nearly 6,000 troops to Afghanistan'. *The Guardian*, p.11.

Norton-Taylor, R. 2009 (16 January). 'Hutton tells NATO allies to 'step up to plate'. over Afghanistan'. *The Guardian*, p.7.

Norton-Taylor, R. 2009b (24 June). 'Insufficient force: British army chiefs are frustrated by Brown's refusal to send more troops to Afghanistan'. *The Guardian*, p.26.

Norton-Taylor, R. and Astill, J. 2003 (5 December). 'Allies at odds over how to fight Afghan drugs boom'. *The Guardian.* http://www.theguardian.com/world/2003/dec/05/afghanistan.foreignpolicy.

Norton-Taylor, R. and Borger, J. 2002 (20 March). 'Commandos embark on uncertain mission: Unease as Hoon says commitment is 'open-ended'. *The Guardian*, p.5.

Norton-Taylor, R. and Macaskill, E. 2002 (30 November). 'Threat of war: New task for British outside Kabul: Military asked to expand its involvement'. *The Guardian*, p.17.

Norton-Taylor, R. and Macaskill, E. 2004 (30 March). 'Britain to send more troops to Afghanistan'. *The Guardian*, p.11.

Norton-Taylor, R. and Vasagar, J. 2006 (10 July). 'Afghanistan: Government commits more troops to offensive'. *The Guardian*, p.18.

PBS. 2007 (23 May). 'Fighting Terrorism in Afghanistan Means Combating Drug Trade'. *PBS Newshour.* http://www.pbs.org/newshour/extra/features/jan-june07/afghanistan_5-23.html.

Porter, A. and Adams, S. 2009 (5 June). 'John Hutton resigns, piling pressure on Gordon Brown'. *The Daily Telegraph.* http://www.telegraph.co.uk/news/politics/gordon-brown/5451115/John-Hutton-resigns-piling-pressure-on-Gordon-Brown.html.

Prince, R. 2008 (10 October). 'Hutton urges allies to stick to promises'. *The Daily Telegraph*, p.18.

Rashid, A. 2005 (16 November). 'Worldstage: Questions the Army must ask before going into Afghanistan'. *The Daily Telegraph*, p.21.

Rayment, S. 2006 (12 February). 'Army rift with No 10 over Afghanistan troops "fiasco"'. *The Sunday Telegraph*, p.1.

Rayment, S. 2009 (12 July). 'Why we can still win this mission: Interview, Bob Ainsworth'. *The Sunday Telegraph*, p.4.

Reeves, P. 2003 (28 February). 'Karzai pleads with US not to abandon Afghanistan'. *The Independent*, p.15.

Sands, S. 2006 (12 October). 'Sir Richard Dannatt: A very honest general'. *The Daily Mail.* http://www.dailymail.co.uk/news/article-410175/Sir-Richard-Dannatt--A-honest-General.html.

Scotsman. 2005 (4 June). 'British to man front line in war against Afghanistan heroin trade'. *The Scotsman.* http://www.scotsman.com/news/uk/british-to-man-front-line-in-war-against-afghanistan-heroin-trade-1-1390697.

Semple, K. and Golden, T. 2007 (7 October). 'U.S. presses again to eradicate Afghan opium poppies'. *The New York Times.* http://www.nytimes.com/2007/10/07/news/07iht-kabul.5.7788904.html?_r=0.

Sengupta, K. and Morris, N. 2009 (18 July). 'Brown snubs Dannatt in talks on reinforcements for Afghanistan'. *The Independent*, p.10.

Sengupta, K. 2007 (5 February). 'Britain hands over Afghanistan to US'. *The Independent*, p.2.

Sengupta, K. and Castle, S. 2006 (30 November) 'Britain to commit extra battalion to NATO'. *The Independent*, p.36.

Sengupta, K. and Taylor, J. 2006 (13 February). 'Into the valley of death: With the Army in a new abuse scandal and no exit strategy from Iraq in sight, UK troops head into another war zone'. *The Independent*, p.19.

Sengupta, K. 2002 (18 January). 'Campaign Against Terrorism: Powell flies to Afghanistan with promises for Karzai but no cash'. *The Independent*, p.6.

Smith, M. 2008 (June 15). 'Brown pulls rank to stop rebel general heading armed forces'. *The Sunday Times*, p.7.

Smith, M. 2006 (23 April). 'Army pleads for more troops after Afghanistan firefight'. *The Sunday Times*, p.4.

Smith, D. 2004 (1 August). 'Britain's war on drugs is naïve, says US'. *The Guardian.* http://www.theguardian.com/world/2004/aug/01/politics.afghanistan.

Spillius, A. 2009 (28 January). 'Goals must be lowered in Afghanistan, says Gates'. *The Daily Telegraph*, p.14.

Starkey, J. 2009 (1 March). 'Step aside, limey, this is how to fight the Taliban'. *The Sunday Times*, p.24.

The Times. 1878 (25 September). 'London, Wednesday, September 25, 1878'. *The Times*, p.9.

Traynor, I. 2009 (9 February). 'US outlines "new realism" in Afghanistan'. *The Guardian*, p.18.

USA Today. 2007 (30 January). 'Dutch troops will not aid destruction of poppy crops in Afghanistan'. *USA Today*. http://usatoday30.usatoday.com/news/world/2007-01-30-dutch-afghanistan_x.htm.

Usborne, D. and Merrick, J. 2009 (5 April). 'NATO at 60: What you need to know'. *The Independent*, p.14.

Walker, T. 2009 (27 January). 'Mandrake: MoD pen mightier than the sword?'. *The Daily Telegraph.* p.8.

Walker, P. 2007 (23 November). 'Defence secretary hits back at admiral's criticism'. *The Guardian.* http://www.theguardian.com/uk/2007/nov/23/iraq.military.

Walsh, D. 2006 (8 July). 'Afghanistan conflict: Desert of death takes its toll on beleaguered troops: British forced to give up hearts and minds mission to stay alive in Afghan outpost'. *The Guardian*, p.18.

Walsh, N.P. and Popalzai, M. 2012 (25 February). '4 killed in Afghanistan amid outrage over Quran burning'. *CNN.* http://edition.cnn.com/2012/02/25/world/asia/afghanistan-burned-qurans.

Watt, N. 2002a (21 March). 'No exit' fear for troops in Afghanistan: War debate: Spectre of Vietnam raised by MPs as Hoon denies the risk of "mission creep"' *The Guardian*, p.12.

Watt, N. 2002b (3 May). 'Troops in Afghanistan: Critics say mission could turn into Britain's Vietnam'. *The Guardian*, p.5.

Watt, N. 2010 (1 July). 'Army cannot leave Afghanistan until job is done, Fox insists'. *The Guardian*, p.13.

Watt, N. 2013 (29 June). 'International forces will provide advice to Afghan military until 2010'. *The Guardian.*

Watt, N. and Wintour, P. 2010 (3 July). 'It's a moral issue... but it's in our national self-interest too'. *The Guardian*, p.13.

Wintour, P. 2010 (26 June). 'Troops out by 2015, says Cameron: Prime minister wants forces to leave Afghanistan before next election'. *The Guardian*, p.1.

Wintour, P. 2009 (30 April). 'Brown redefines British goals in Afghanistan: Aim reduced to setting up a functioning state'. *The Guardian*, p.8.

Wintour, P. 2006 (8 July). 'Interview: Des Browne: 'No one ever suggested it was going to be easy'. *The Guardian*, p.18.

Wintour, P. 2004 (3 May). 'Battle begins to stem Afghan opium harvest: Britain must adopt more aggressive tactics to prevent record year for poppy growers, US warns'. *The Guardian*, p.12.

Wright, O. 2014 (27 May). "Costly failures': Wars in Iraq and Afghanistan cost UK taxpayers £30bn'. *The Independent.* http://www.independent.co.uk/news/uk/politics/costly-failures-wars-in-iraq-and-afghanistan-cost-uk-taxpayers-30bn-9442640.html.

Zoroya, G. and Leinwand, D. 2004 (26 October) 'Rise of drug threat to Afghanistan's security'. *USA Today.* http://usatoday30.usatoday.com/news/world/2004-10-26-opium-afghanistan_x.htm.

Parliamentary Statements (Hansard).

Ainsworth, B. (2008e) 29 October. Hansard WH Debate. Vol.481, Col.284WH.
Blair, T. (2001a) 14 September. Hansard HC Debate. Vol.372, Col.604-607.
Blair, T. (2001b) 4 October. Hansard HC Debate. Vol.372, Col.673.
Blair, T. (2001d) 14 November. Hansard HC Debate. Vol.374, Col.864.
Blair, T. (2002c) 24 September. Hansard HC Debate. Vol.390, Col.1-7.
Blair, T. (2003a) 18 March. Hansard HC Debate. Vol.401, Col.761, 768.
Blair, T. (2004b) 19 April. Hansard HC Debate. Vol.420, Col.35.
Blair, T. (2004e) 30 June. Hansard HC Debate. Vol.424, Col.287.
Blair, T. (2004f) 8 September. Hansard HC Debate. Vol.424, Col.714.
Brown, G. (2007d) 12 December. Hansard HC Debate. Vol.469, Col.303-306, 318.
Brown, G. (2008d) 2 July. Hansard HC Debate. Vol.478, Col.854.
Brown, G. (2009b) 29 April. Hansard HC Debate. Vol.491, Col.869-873.
Brown, G. (2009c) 14 October. Hansard HC Debate. Vol.497, Col.313.
Browne D. (2007a) 22 January. Hansard HC Debate. Vol.455, Col.1134.
Browne D. (2008b) 16 June. Hansard HC Debate. Vol.477, Col.681.
Cameron, D. (2007e) 12 December. Hansard HC Debate. Vol.469, Col.307.
Cameron, D. (2010c) 14 June. Hansard HC Debate. Vol.511, Col.604.
Cameron, D. (2011) 6 July. Hansard HC Debate. Vol.530, Col.1511-1513.
Cameron, D. (2013a) 29 August. Hansard HC Debate. Vol.566, Col.1425,1437.
Cameron, D. (2014) 24 September. Hansard HC Debate. Vol.585, Col.1255-1256.
Flynn, P. (2006c) 9 February. Hansard HC Debate. Vol.442, Col.1016.
Flynn, P. (2007b) 11 July. Hansard WH Debate. Vol.462, Col.423WH.
Fox, L. (2008a) 16 June. Hansard HC Debate. Vol.477, Col.679, 680.
Fox, L. (2009c) 15 October. Hansard HC Debate. Vol.497, Col.477.
Fox, L. (2010d) 9 September. Hansard HC Debate. Vol.515, Col.508, 512.
Fox, L. (2013c) 29 August. Hansard HC Debate. Vol.566, Col.1452-1453.
Grey, H.G. (1878) 10 December. Hansard HL Debate. Vol.243, Col.414.
Harcourt, W. (1880) 11 February. Hansard HC Debate. Vol.250, Col.468.
Harvey, N. (2008c) 16 June. Hansard HC Debate. Vol.477, Col.682.
Hoon, G. (2002d) 17 October. Hansard HC Debate. Vol.390, Col.505-506.
Hutton, J. (2008e) 30 October. Hansard HC Debate. Vol.481, Col.1073.
Ingram, A. (2005a) 4 July. Hansard HC Debate. Vol.436, Col.17.
Leigh, E. (2013d) 29 August. Hansard HC Debate. Vol.566, Col.1521.
Mccarthy-Fry, S. (2007c) 12 November. Hansard HC Debate. Vol.467, Col.470.
Miliband, D. (2009a) 5 February. Hansard HC Debate. Vol.487, Col.1034.
Miliband, D. (2010a) 14 January. Hansard HC Debate. Vol.503, Col.878.
Miliband, D. (2010b) 1 February. Hansard HC Debate. Vol.505, Col.33.

Mordaunt, P. (2013b) 29 August. Hansard HC Debate. Vol.566, Col.1437,1447.
Osborne, G. (2004a) 2 March. Hansard HC Debate. Vol.418, Col.738.
Rammell, B. (2004d) 9 June. Hansard WH Debate. Vol.422, Col.122WH-126WH.
Reid, J. (2005b) 17 October. Hansard HC Debate. Vol.437, Col.618.
Reid, J. (2005c) 14 November. Hansard HC Debate. Vol.439, Col.679-681.
Reid, J. (2005d) 12 December. Hansard HC Debate. Vol.440, Col.1093.
Reid, J. (2006a) 26 January. Hansard HC Debate. Vol.441, Col.1529-1532.
Reid, J. (2006d) 4 May. Hansard WH Debate. Vol.445, Col.1812W.
Short, C. (2001c) 8 October. Hansard HC Debate. Vol.372, Col.899.
Short, C. (2002a) 28 January. Hansard HC Debate. Vol.379, Col. 28, 34.
Stanley, J. (2004c) 20 May. Hansard WH Debate. Vol.421, Col.320WH.
Straw, J. (2002b) 11 July. Hansard HC Debate. Vol.388, Col.1132.
Straw, J. (2003b) 27 November. Hansard HC Debate. Vol.415, Col.144.

Speeches and Press Conferences.

Ban, K.M. 2007. 'Remarks by United Nations Secretary-General Ban Ki-Moon at a joint press conference with President of Afghanistan Hamid Karzai and Italian Foreign Minister Massimo D'Alema'. *Federal News Service.*
BBC News. 2010. 'Second prime ministerial debate, 22 April 2010'. *BBC.* http://news.bbc.co.uk/1/shared/bsp/hi/pdfs/23_04_10_seconddebate.pdf.
Blair, T. 1995. 'Leader's speech, Brighton 1995'. http://www.britishpoliticalspeech.org/speech-archive.htm?speech=201.
Blair, T. 1999. (22 April). 'The Blair Doctrine'. *PBS.* http://www.pbs.org/newshour/bb/international/jan-june99/blair_doctrine4-23.html.
Blair, T. 2001a. (11 September). 'Blair's statement in full'. *BBC.* http://news.bbc.co.uk/1/hi/uk_politics/1538551.stm.
Blair, T. 2001b. (2 October). 'Leader's speech, Brighton 2001'. http://www.britishpoliticalspeech.org/speech-archive.htm?speech=186.
Blair, T. 2002a (8 April). 'Full text of Tony Blair's speech in Texas'. *The Guardian.* http://www.theguardian.com/politics/2002/apr/08/foreignpolicy.iraq.
Blair, T. 2002a (10 September). 'Speech to TUC conference, Blackpool 2002'. http://www.britishpoliticalspeech.org/speech-archive.htm?speech=284.
Blair, T. 2002b (1 October). 'Full text of Tony Blair's speech (1)' *The Guardian.* http://www.theguardian.com/politics/2002/oct/01/labourconference.labour14.
Blair, T. 2003 (17 July). 'Speech to the US Congress, Washington DC 2003'. http://www.britishpoliticalspeech.org/speech-archive.htm?speech=285.
Blair, T. 2004 (28 September). 'Leader's speech, Brighton 2004'. http://www.britishpoliticalspeech.org/speech-archive.htm?speech=183.
Blair, T. 2005 (27 September). 'Leader's speech, Brighton 2005'. http://www.britishpoliticalspeech.org/speech-archive.htm?speech=182.
Blair, T. 2006 (21 March). 'Tony Blair's speech to the Foreign Policy Centre'. *The Guardian.* http://www.theguardian.com/politics/2006/mar/21/iraq.iraq1.

Blair, T. 2006b (26 September). 'Tony Blair's speech'. *The Guardian*. http://www. theguardian.com/politics/2006/sep/26/labourconference.labour3.

Brown, G. 2005 (26 September). 'Chancellor's speech, Brighton 2005'. http:// www.britishpoliticalspeech.org/speech-archive.htm?speech=276.

Cameron, D. 2010 (6 October). 'David Cameron's Conservative conference speech in full'. *The Daily Telegraph*. http://www.telegraph.co.uk/news/politics/david-cameron/8046342/David-Camerons-Conservative-conference-speech-in-full. html.

Charles, R. 2004 (1 April). 'Afghanistan: Are the British Counternarcotics Efforts Going Wobbly?'. Subcommittee on Criminal Justice, Drug Policy and Human Resources: House of Representative, 118th Congress, Second Session. http:// www.gpo.gov/fdsys/pkg/CHRG-108hhrg96745/html/CHRG-108hhrg96745. htm.

Cook, R. 1997 (12 May). 'Robin Cook's speech on the government's ethical foreign policy'. *The Guardian*. http://www.theguardian.com/world/1997/ may/12/indonesia.ethicalforeignpolicy.

Fox, L. 2010b (11 September). 'The Strategy for Afghanistan'. http://www.mod. uk/DefenceInternet/AboutDefence/People/Speeches/SofS/20100911TheStrate gyForAfghanistan.htm.

Fox, L. 2010c (22 November). 'Address to the Vivekananda International Foundation'.   http://www.mod.uk/DefenceInternet/AboutDefence/People/ Speeches/SofS/20101122AddressToTheVivekanandaInternationalFoundati on.htm.

Fox, L. 2011b (30 August). 'Partnership with Purpose: Multi-Layered Security in the 21st Century'. http://www.mod.uk/ DefenceInternet/AboutDefence/People/Speeches/SofS/20110830Part nershipWithPurposeMultilayeredSecurityInThe21stCentury.htm.

Hammond, P. 2011 (8 December). 'Delivering on the Frontline: Operational Success and Sustainable Armed Forces'. *RUSI*. https://www.rusi.org/events/ past/www/ref:E4EBAA415CBC7A.

Hammond, P. 2012 (29 March). 'Defence Secretary visits Afghanistan and agrees new officer academy'. *Ministry of Defence*. https://www.gov.uk/government/ news/defence-secretary-visits-afghanistan-and-agrees-new-officer-academy.

Harvey, N. 2010 (14 December). '2010/12/14 – Denmark-UK Defence Co-operation'. https://www.gov.uk/government/speeches/2010-12-14-denmark-uk-defence-co-operation.

Hutton, J. 2008b (11 November). 'Speech by the Rt Hon John Hutton, Secretary of State for Defence 11th November 2008'. *IISS*. http://www.iiss.org/recent-key-addresses/john-hutton-address.

Miliband, D. 2007a (19 July). 'New Diplomacy: Challenges for Foreign Policy'. http://davidmiliband.net/speech/new-diplomacy-challenges-for-foreign-policy.

Miliband, D. 2007b (5 September). 'Shared Values and Shared Future: The Importance of Turkey To Our Common Future'. http://davidmiliband.net/

speech/shared-values-and-shared-future-the-importance-of-turkey-to-our-common-future.

Miliband, D. 2007c (11 November). 'Launch of the Hajj Delegation'. http://davidmiliband.net/speech/launch-of-the-hajj-delegation-2.

Miliband, D. 2008a (19 January). 'Change The World'. http://davidmiliband.net/speech/change-the-world.

Miliband, D. 2008b (25 March). 'FCO's 2007 Human Rights Report'. http://davidmiliband.net/speech/fcos-2007-human-rights-report.

Miliband, D. 2008c (21 May). 'Dilemmas of Democracy: Afghanistan and Pakistan'. http://davidmiliband.net/speech/dilemmas-of-democracy-afghanistan-and-pakistan.

Miliband, D. 2008d (7 July). 'Freedom and Responsibility: New Challenges in Africa'. http://davidmiliband.net/speech/freedom-and-responsibility-new-challenges-in-africa.

Miliband, D. 2008e (21 November). 'Foundations Of Freedom: The Promise Of The New Multilateralism'. http://Davidmiliband.Net/Speech/Foundations-Of-Freedom-The-Promise-Of-The-New-Multilateralism.

Miliband, D. 2009c (27 July). 'NATO's mission in Afghanistan: The political strategy'. http://davidmilibandarchive.blogspot.co.uk/2013/09/natos-mission-in-afghanistan-political.html.

Miliband, D. 2009d (17 November). '"The War in Afghanistan: How a Political Surge Can Work": Speech by Rt Hon David Miliband MP, Foreign Secretary'. *NATO.* www.nato-pa.int/Docdownload.asp?ID=0B69DCCA5F050C0E0014.

Obama, B. 2009 (27 March). 'Remarks by the President on a New Strategy for Afghanistan and Pakistan'. *The White House.* http://www.whitehouse.gov/the-press-office/remarks-president-a-new-strategy-afghanistan-and-pakistan.

Obama, B. 2011 (22 June). 'Remarks by the President on the Way Forward in Afghanistan'. *The White House.* http://www.whitehouse.gov/the-press-office/2011/06/22/remarks-president-way-forward-afghanistan.

Obama, B. 2014 (10 September). 'Statement by the President on ISIL'. *The White House.* http://www.whitehouse.gov/the-press-office/2014/09/10/statement-president-isil-1.

O'Loughlin, B. 2012. 'Can Strategic Narrative Be Effective? Infrastructure, Intention, Experience – Ben O'Loughlin'. *YouTube.* http://www.youtube.com/watch?v=_11vprLADZY.

Reid, J. 2005 (10 June). 'Speech by Dr. John Reid, UK Secretary of State for Defence'. *NATO.* http://www.nato.int/docu/speech/2005/s050610h.htm.

Reid, J. 2006 (9 February). 'Secretary of Defense Donald Rumsfeld and UK Minister of Defense John Reid Press Stakeout at the NATO Defense Ministerial'. *US Department of Defense.* http://www.defense.gov/Transcripts/Transcript.aspx?TranscriptID=936.

Rizzo, A.M. 2003 (11 August). 'Speech by the Deputy Secretary, Alessandro Minuto Rizzo at the ISAF Assumption of Command Ceremony'. *NATO.* http://www.nato.int/docu/speech/2003/s030811a.htm.

Robertson, G. 2003 (29 September). 'Speech by NATO Secretary General, Lord Robertson'. *NATO.* http://www.nato.int/docu/speech/2003/s030929a.htm.

Stirrup, J. 2009 (3 December). 'RUSI Christmas Lecture'. *Ministry of Defence.* http://www.mod.uk/DefenceInternet/AboutDefence/People/Speeches/ChiefSt aff/20091203RusiChristmasLecture.htm.

Public Interviews.

Ainsworth, B. 2010a (11 June). 'Bob Ainsworth backs keeping UK troops in Afghanistan'. *BBC: The Daily Politics.* http://news.bbc.co.uk/1/hi/ programmes/the_daily_politics/8735405.stm.

Ainsworth, B. 2010b (9 September). 'Bob Ainsworth on British troops in Afghanistan'. *BBC: The Daily Politics.* http://www.bbc.co.uk/news/uk-politics-11247135.

Alexander, D. 2008 (3 February). 'Afghanistan: "A long struggle"'. *BBC: The Andrew Marr Show.* http://news.bbc.co.uk/1/hi/programmes/andrew_marr_ show/7224856.stm.

Benn, H. 2006 (19 November). 'International Aid'. *BBC: Sunday AM.* http://news. bbc.co.uk/1/hi/programmes/sunday_am/6163046.stm.

Blair, T. 2007 (28 January). 'Blair: "I'm going to finish what I started"'. *BBC: The Politics Show.* http://news.bbc.co.uk/1/hi/programmes/politics_ show/6293605.stm.

Cameron, D. 2006 (1 October). 'Tory tax cuts ruled out?'. *BBC: The Andrew Marr Show.* http://news.bbc.co.uk/1/hi/programmes/sunday_am/5396492.stm.

Cameron, D. 2013 (16 December). 'BBC News Afghanistan mission accomplished, says David Cameron'. *BBC.* https://www.youtube.com/watch?v=dIozyL55c2c.

Fox, L. 2007 (4 March). 'NATO questioned'. *BBC: Sunday AM.* http://news.bbc. co.uk/1/hi/programmes/sunday_am/6417101.stm.

Fox, L. 2009 (18 October). 'Met Office sale potential – Fox'. *BBC: The Andrew Marr Show.* http://news.bbc.co.uk/1/hi/programmes/andrew_marr_ show/8313106.stm.

Fox, L. 2010 (28 January). 'Fox, Hutton and Frogh on Afghanistan and the Taliban – 28/01/2010 The Daily Politics'. *BBC: The Daily Politics.* http://news.bbc. co.uk/1/hi/programmes/the_daily_politics/8485318.stm.

Fox, L. 2011a (27 March). 'Transcript of Liam Fox interview'. *BBC: The Andrew Marr Show.* http://news.bbc.co.uk/1/hi/programmes/andrew_marr_ show/9437722.stm.

Hutton, J. 2009 (15 March). 'John Hutton interview transcript'. *BBC: The Politics Show.* http://news.bbc.co.uk/1/hi/programmes/politics_show/7932367.stm.

Hutton, J. 2008a (26 October). 'Hutton: We could be there for decades'. *BBC: The Politics Show.* http://news.bbc.co.uk/1/hi/programmes/politics_ show/7680732.stm.

Karzai, H. 2005 (22 May). 'CNN Late Edition with Wolf Blitzer: Interview with Hamid Karzai'. *CNN.* http://transcripts.cnn.com/TRANSCRIPTS/0505/22/le.01.html.

Miliband, D. 2007 (15 July). 'Labour foreign policy'. *BBC: The Andrew Marr Show.* http://news.bbc.co.uk/1/hi/programmes/sunday_am/6899510.stm.

Miliband, D. 2009a (5 July). '"Cold anger" at Iran'. *BBC: The Andrew Marr Show.* http://news.bbc.co.uk/1/hi/programmes/andrew_marr_show/8134940.stm.

Miliband, D. 2009b (25 October). 'Miliband defiant on Blair EU role'. *BBC: The Andrew Marr Show.* http://news.bbc.co.uk/1/hi/programmes/andrew_marr_show/8324790.stm.

Rammell, B. 2009a (1 October). 'Daily Politics – Bill Rammell on Afghan campaign'. *BBC: The Daily Politics.* http://news.bbc.co.uk/1/hi/programmes/the_daily_politics/8285035.stm.

Rammell, B. 2009b (11 December). 'HardTalk & Bill Rammell on UK Role in Afghanistan 1 of 3 – BBC News Interview'. *BBC: HardTalk.* http://www.youtube.com/watch?v=b9EYBb9ayG0&list=PLF0975E652EF0797D&index=22&feature=plpp_video.

Rasmussen, A.F. 2012 (3 May). 'Boulton & Co 03.05.12 Anders Fogh Rasmussen, NATO Secretary-General'. *Sky News.* http://skynews.skypressoffice.co.uk/newstranscripts/boulton-co-030512-anders-fogh-rasmussen-nato-secretary-general.

Reid, J. 2006a (19 February). 'Be slow to condemn'. *BBC Sunday AM.* http://news.bbc.co.uk/1/hi/programmes/sunday_am/4729120.stm.

Scheffer, J. 2004 (22 October). 'Press Point with NATO Secretary General, Jaap de Hoop Scheffer and the Prime Minister of Hungary, Mr. Ferenc Gyurcsany'. *NATO.* http://www.nato.int/docu/speech/2004/s041022a.htm.

Straw, J. 2002 (24 March). 'BBC Breakfast with Frost Interview: Jack Straw MP Foreign Secretary March 24th, 2002'. *BBC.* http://news.bbc.co.uk/1/hi/programmes/breakfast_with_frost/1890782.stm.

Other Publications.

Asia Foundation. 2011. 'Afghanistan in 2011: A Survey of the Afghan People'. *The Asia Foundation.*

Butler, Lord. 2004. 'Review of Intelligence on Weapons of Mass Destruction: Report of a Committee of Privy Counsellors'. *House of Commons.*

Cabinet Office. 2008. 'The National Security Strategy of the United Kingdom: Security in an interdependent world'. *HM Government.*

Cabinet Office. 2010a. 'A Strong Britain in an Age of Uncertainty: The National Security Strategy'. *HM Government.*

Cabinet Office. 2010b. 'Securing Britain in an Age of Uncertainty: The Strategic Defence and Security Review'. *HM Government.*

Development, Concepts and Doctrine Centre. 2002. *Joint Warfare Publication 3-80: Information Operations*. Ministry of Defence: Shrivenham..

Development, Concepts and Doctrine Centre. 2009. *Joint Doctrine Publication 3-40: Security and Stabilisation*. Ministry of Defence: Shrivenham.

Development, Concepts and Doctrine Centre. 2010a. *The Future Character of Conflict*. Ministry of Defence: Shrivenham.

Development, Concepts and Doctrine Centre. 2010b. *Joint Doctrine Publication 04: Understanding*. Ministry of Defence: Shrivenham.

Development, Concepts and Doctrine Centre. 2011a. *Joint Doctrine Publication 0-01: UK Defence Doctrine*. Ministry of Defence: Shrivenham.

Development, Concepts and Doctrine Centre. 2011b. *Joint Doctrine Note 1/11: Strategic Communication*. Ministry of Defence: Shrivenham.

Development, Concepts and Doctrine Centre. 2012a. *Joint Doctrine Note 1/12: Strategic Communication*. Ministry of Defence: Shrivenham.

Government of Germany. 2004 (1 April). 'The Berlin Declaration on Counter-Narcotics'. *AG Friedensforschung*. http://www.ag-friedensforschung.de/regionen/Afghanistan/berlin-antidrogen.html.

HM Government. 1998. 'Tackling Drugs to Build a Better Britain: The Government's Ten-Year Strategy for Tackling Drugs Misuse'. https://www.gov.uk/government/uploads/system/uploads/attachment_data/file/259785/3945.pdf.

House of Commons Defence Committee. 2006. 'The UK deployment to Afghanistan: Fifth Report of Session 2005-06'. *Parliament*.

House of Commons Defence Committee. 2007. 'UK operations in Afghanistan: Thirteenth Report of Session 2006–07'. *Parliament*.

House of Commons Defence And Foreign Affairs Committee. 2008. 'Iraq and Afghanistan'. *Parliament*. http://www.publications.parliament.uk/pa/cm200708/cmselect/cmdfence/1145/8102801.htm.

House Of Commons Defence Committee. 2009. 'The Comprehensive Approach: The point of war is not just to win but to make a better peace: Seventh Report of Session 2009–10'. *Parliament*.

House of Commons Defence Committee. 2013. 'Securing the Future of Afghanistan: Tenth Report of Session 2012–13'. *Parliament*.

House of Commons Public Administration Select Committee. 2010. 'Who does UK National Strategy: First Report of Session 2010–11'. *Parliament*.

House of Lords Select Committee on Communications. 2008. 'Government Communications: 1st Report of Session 2008–09'. *Parliament*.

ICM. 2006. 'Afghanistan Survey'. http://www.icmresearch.com/pdfs/2006_september_bbc_afghanistan_poll.pdf.

ICM. 2009. 'ICM Poll for the BBC/The Guardian'. http://www.icmresearch.com/pdfs/2009_july_guardiian_bbc_afghanistan_poll.pdf.

Iraq Inquiry. 2010a (3 February) Oral Evidence given by John Reid. http://www.iraqinquiry.org.uk/media/45011/20100203am-reid-final.pdf.

Iraq Inquiry. 2010b (5 March) Oral Evidence given by Gordon Brown. http://www.iraqinquiry.org.uk/media/45558/100305-brown-final.pdf.

Iraq Inquiry. 2010c (28 July). Oral Evidence given by General Sir Richard Dannatt. http://www.iraqinquiry.org.uk/media/53218/Dannatt%202010-07-28%20S1.pdf.

Islamic Republic of Afghanistan. 2003. *National Drug Control Strategy.*

The London Conference on Afghanistan. 2006. 'The Afghanistan Compact'. *NATO.* http://www.nato.int/isaf/docu/epub/pdf/afghanistan_compact.pdf.

Mcchrystal, S.A. 2009. 'Commander's Initial Assessment'. *International Security Assistance Force.*

UNODC. 2004a (6 February). 'United Nations Office Sees "Critical Decisions" For Counter-Narcotic Efforts in Afghanistan'. *UNODC.* http://www.unodc.org/unodc/en/press/releases/press_release_2004-02-06_2.html.

UNODC. 2004b (9 February). 'International Counter Narcotics Conference on Afghanistan 8–9 February 2004'. https://www.unodc.org/pdf/afg/afg_intl_counter_narcotics_conf_2004.pdf.

UNODC. 2004c. 'Afghanistan Opium Survey 2004: November 2004'. *United Nations Office on Drugs and Crime.*

UNODC. 2005. 'Afghanistan Opium Survey 2005: November 2005'. *United Nations Office on Drugs and Crime.*

UNODC. 2006. 'Afghanistan Opium Survey 2006: October 2006'. *United Nations Office on Drugs and Crime.*

UNODC. 2007. 'Afghanistan Opium Survey 2007: Executive Summary: August 2007'. *United Nations Office on Drugs and Crime.*

UNODC. 2009. 'Afghanistan Opium Survey 2009: Summary Findings: September 2009'. *United Nations Office on Drugs and Crime.*

UNODC. 2011. 'Afghanistan Opium Survey 2011: December 2011'. *United Nations Office on Drugs and Crime.*

UNODC. 2013. 'Afghanistan Opium Survey 2013: Summary findings: November 2013'. *United Nations Office on Drugs and Crime.*

US Department of Defense. 2011 (10 June). 'Gates: NATO has become a two-tiered alliance'. *United States Department of Defense.* http://www.defense.gov/news/newsarticle.aspx%3Fid%3D64268.

The White House. 2003. *National Drug Control Strategy: Update.*

Index

16 Air Assault Brigade 25, 100

Afghan Compact (2006) 40, 43, 44, 60,
    72, 101, 136
Afghan Interim Authority 48, 119
Afghan National Development Strategy
    (ANDS) 61, 82
Afghan National Drug Control Strategy 91
Afghan National Security Forces (ANSF)
    1, 2, 33, 64, 141
Ainsworth, Bob 78, 80, 134, 138
Akhundzada, Sher Mohamed 101
Alexander, Douglas 29, 66, 68, 105–6,
    129–30
Arab Spring 83, 167
Armed Forces (UK) 13, 16, 21, 23, 24–5,
    31, 35, 60, 61, 67–9, 74–5, 77, 116,
    134
Army (UK) 22, 23–5, 34, 62, 68, 69, 73,
    74, 77, 102
Assad, al-, Bashar 83, 159
Australia 59

Bahrain 83
Ban Ki-Moon 65
Barkawi, Tarak 46, 70
Benn, Hilary 29, 62
Berlin Conference (2004) 52, 94–5, 122
Betz, David 16, 25, 29, 33, 57
Blackpool Trade Union Congress
    Conference (2002) 89
Blair, Tony; Government 13, 18, 23, 24,
    31, 32, 39, 41–2, 44–7, 48–55, 57,
    59–67, 71, 73, 75, 79, 83, 85, 88–9,
    91, 92–3, 97, 98, 99, 100–105,
    111–12, 116–30, 136, 138, 140,
    143–4, 150–51, 152, 154, 158, 159,
    160, 168
Bonn Agreement (2001) 40, 47–8, 101
Børgesen, Berit 143, 149

Boyce, Michael Lord 67
Brighton Labour Party Conference (2001)
    89, 117
Brighton Labour Party Conference (2005)
    124–5
Brown, Gordon; Government 14, 16,
    24–5, 29, 32, 44, 63–79, 81,
    84, 85, 104–10, 116–17, 124–5,
    128–36, 138–40, 141–2, 168
Browne, Des 24–5, 29, 31, 61–2, 63, 66,
    68, 103, 129–30, 168
Bucharest NATO Summit (2008) 67, 105
Burke, Jason 49
Bush, George W.; Administration 17, 23,
    39, 45–6, 50, 54, 65, 66, 70, 71,
    73, 96, 104, 108, 119, 121, 128,
    129, 143
Butler, Brig Ed 102
Butler Report (2004) 55

Cabinet 31–2, 34, 61, 65, 73, 75, 79, 91,
    109, 110, 117, 120, 129, 130, 139,
    142
Cairns, David 95
Cameron, David; Government 17, 44, 64,
    78–85, 109, 110, 115–17, 127, 131,
    139, 141–3, 159–61, 168
Canada 59, 101, 134
Cavagnari, Sir Charles 20
Cavanagh, Matt 43, 55–6, 135–6, 140
Charles, Robert 93–4
Chicago NATO Summit (2012) 80–81
Clarke, Michael 99–101
Clausewitz, Carl von 163–5, 167
Colley, Paul 155
Collins, Colonel Tim 62
'Comprehensive Approach' (CA) 10,
    14, 23, 26, 42, 59, 64–6, 68, 81,
    82, 84, 88, 96, 98, 109, 112, 113,
    129, 130, 152, 164

Cook, Robin 41, 53
Cormack, Anthony 16, 25, 29, 33, 57
Cornish, Paul 9, 21–2, 34, 152
Costa, Antonio Maria 93, 108
Cowper-Coles, Sherard 55

Daily Mail, The 24
Dalacoura, Katerina 83
Dannatt, General Sir Richard 24, 29, 67,
    69, 74, 76–7
Daud, Mohamed 101
Denmark 59, 143
Denny, Col James 57
Department for International Development
    (DfID) 23, 25–6, 29, 31, 33–4, 43,
    66, 68, 69, 113, 129
Dimitriu, George 166
Disraeli, Benjamin 18–20
Dorman, Andrew 13, 21, 22, 80
Dostum, Abdul Rashid 47–8
Drug Enforcement Agency (DEA) (US)
    104

Edmunds, Timothy 10, 13–14, 16–17,
    21–2, 26, 27, 29
Egypt 83
Estonia 59

Fedotov, Yury 110
Felbab-Brown, Vanda 90
Flynn, Paul 104
Foreign and Commonwealth Office (FCO)
    23, 25–6, 31, 33, 34, 43, 63, 66, 68,
    90, 91, 92, 94, 95, 96, 100, 102,
    106, 113, 129, 135
Fox, Liam 24, 31, 44, 59, 67, 69, 78–9,
    82–3, 109, 128, 131, 136–7, 138,
    160
France 50, 53, 134
Frost, David 119
Fry, Sir Robert 4, 5, 42

Gaddafi, Muammar 83–4
Gates, Robert 71, 74
Germany 50, 53, 89, 128, 134
Gladstone, William 18, 20
Gowing, Nik 19
Gray, Colin 15, 17, 22, 155

Grey, Earl 19–20
Griffin, Stuart 168
Guardian, The 59, 102, 139
Guthrie, Charles Lord 60

Hadir, Abdul 48
Hain, Peter 118
Hames, Tim 72
Hammond, Philip 142
Harcourt, William 20
Hartington, The Marquess of 18
Heathcoat-Amory, David 132–3
Holbrooke, Richard 108
Hoon, Geoff 24, 48–9, 120
Hoop Scheffer, Jaap de-, 123–4
House of Commons Defence Committee
    101, 132–3
House of Commons Foreign Affairs
    Select Committee 52–3, 107, 111,
    139–40
House of Commons Public Administration
    Committee 4
Howells, Kim 63–4
Hussein, Saddam 50, 120–21
Hutton, John 30–32, 68–9, 72, 74–5, 78,
    82, 129–32, 134, 146, 168

Inge, Field Marshal Peter 3
Ingram, Adam 57
International Security Assistance Force
    (ISAF) 1, 2, 6, 10, 15, 17, 21, 23,
    33, 38, 39, 40, 44, 47, 48, 50, 51,
    52, 53, 54, 57, 59, 60, 63, 74, 79,
    84, 93, 95, 98, 100, 101, 105, 107,
    108, 110, 119, 121, 122, 123, 124,
    125, 126, 128, 133, 135, 140, 141
Iraq 13, 17, 23, 24, 29, 30, 32, 38, 43,
    45, 50, 51–3, 54–5, 60, 62, 63, 67,
    70, 71, 72, 75, 84, 99, 107, 112,
    120–22, 124, 127, 128–9, 133, 135,
    145, 157, 159, 160
Islamic State in Iraq and the Levant (ISIL)
    84, 157–60
Istanbul NATO Summit (2004) 53–5, 122
Italy 89, 134

Jackson, General Sir Mike 67, 103
Janis, Irving 55

Japan 89
Jenkins, Simon 102

Kant, Immanuel 45
Karzai, Hamid 50, 51, 53, 60, 64, 78, 93,
    96, 97, 104, 105
Khan, Yakub 20
King, Anthony 7, 10, 16, 17, 26, 28, 142,
    155
Kirkup, James 32
Kreps, Sarah 134

Laden, bin-, Osama 39, 70, 80, 119, 141
Lawford, Col Huw 57
Layton, Peter 155
Lemahieu, Jean-Luc 111
Libya 13, 83–4, 157
Lindley-French, Julian 152
Lisbon NATO Summit (2010) 39, 79, 80,
    139, 140, 146, 158
Lodge, Guy 31, 129–30
London Conference (2006) 60, 63, 79, 82,
    101–2
London Conference (2010) 108–9
Lorimer, Brig John 63

Mackay, Andrew 26
MacShane, Denis 78
McChrystal, General Stanley 79, 108, 166
Malloch-Brown, Lord Mark 106–7, 135
Manchester Labour Party Conference
    (2006) 127
Massoud, Ahmed Zia 104
Metz, Steven 70
Miliband, David 31, 33, 66, 68, 72, 77, 83,
    108, 129–30, 132–3, 134, 136, 159
Ministry of Defence (MoD) 8, 25, 26, 30,
    31–2, 34, 43, 46, 57, 60, 68, 69, 70,
    72, 73, 79, 98, 101, 103, 130, 134,
    146, 152, 164
Monten, Jonathan 45

National Security Council (UK) 34, 79
Netherlands, The 79, 101, 105, 128
New Labour 39, 46, 87, 88, 117, 122
Newton, Paul 155
North Atlantic Treaty Organisation
    (NATO) 4, 5, 6–7, 9, 14–15, 17,
    21, 31, 32, 36, 37, 38, 40, 41, 43,
    47, 50, 51, 52–7, 59, 62, 64, 67, 68,
    69, 71, 73–4, 79, 80, 82–5, 93, 98,
    99, 101, 105, 107, 110, 111, 115,
    118, 121, 122–4, 126, 128, 132,
    134, 136, 139–40, 142, 143, 145–7,
    149, 153, 156–9, 161
Northern Alliance 46, 47, 48
Norton-Taylor, Richard 72

Obama, Barack; Administration 17, 32–3,
    39, 70–75, 79, 80, 107, 108, 111,
    117, 133–4, 135, 140, 141, 161
O'Brien, Mike 91
Operation Anaconda 141
Operation Enduring Freedom (OEF) 1, 21,
    47, 95, 126
Operation Veritas 119
Osborne, George 94

Pakistan 1, 64, 74, 108, 119, 124, 133,
    134–6, 141, 142, 158
Porter, Patrick 155
Porter, Robin 16–17
Post Conflict Reconstruction Unit (PCRU)
    25–6, 61
Provincial Reconstruction Team (PRT) 50,
    53, 102

Qaeda, al-, 2, 45, 46, 49, 65, 87, 89, 92,
    93, 94, 106, 109, 124, 126, 133,
    134, 135, 141

Rabbani, Burhanuddin 48
Rammell, Bill 77–8, 94–6, 134, 136, 138
Rasmussen, Anders Fogh 142–3
Reid, John 24, 29, 43, 55, 57, 59, 60–61,
    98, 100–106, 111, 126, 127, 135,
    168
Richards, General Sir David 100, 102,
    134–5, 142
Ricks, Thomas 70
Riga NATO Summit (2007) 62
Ringsmose, Jens 143, 149
Rizzo, Alessandro Minuto 123
Roberts, Major General Frederick 21
Robertson, Lord George 123
Roenfeldt, Carsten 163, 166

Rogers, James 155
Royal Air Force 22, 34
Royal Marines 22, 48, 119
Royal Navy 22, 34
Rumsfeld, Donald 49, 50, 126
Russia 18, 19, 110

Scotsman, The 98
Second Afghan War 10, 18–21
Secret Intelligence Service (SIS) 90
Seldon, Anthony 31, 129
Senate Foreign Relations Committee 50
Sharpe, Andrew 155
Short, Clare 25, 26, 27, 89–90, 91
Simon, Luis 155
Simpson, Emile 163–6
Smith, Rupert 9, 22
Somalia 13, 74, 136
Spain 59
Special Air Service (SAS) 119
Stirrup, Jock 4, 5, 42, 137
Strachan, Hew 164
Strategic Communication (SC) 3, 6, 8–11,
    21, 30–34, 35–6, 38, 44, 59, 70,
    73–4, 81–2, 109, 113, 115, 116,
    134, 137, 139, 146, 149, 151–4,
    156–69
Straw, Jack 49, 52, 91, 119
Syria 13, 83–4, 159–61

Taliban 2, 46, 48, 49, 50, 52, 62, 63, 72,
    75, 79, 84, 87, 89, 93, 100, 101,
    102–4, 106, 110–11, 130, 133,
    134, 136
Task Force Helmand 42, 43, 102, 106, 126

Telegraph, The 32, 73, 75, 104, 137
Times, The 19, 25, 74, 98
Tokyo Conference (2002) 47, 55, 89, 90,
    92, 101
Toronto G8 Summit (2010) 79, 139

United Nations 15, 107, 121, 127
United Nations Office on Drugs and Crime
    (UNODC) 91, 93, 96, 97, 108,
    110–11
United Nations Refugee Agency 52
United Nations Security Council 15, 51
United States 1, 10, 13, 14, 15–17, 21, 27,
    32, 34, 38–9, 41, 46, 47, 48, 50, 51,
    53, 56, 59, 63, 65, 69, 80, 83, 85,
    87, 88, 89, 91, 92, 93, 94, 95, 96,
    99, 103–6, 107, 108, 118, 120–21,
    129, 130, 133, 134, 135, 138, 141,
    142, 144, 151, 152, 154–6
United States Department of State 92, 95,
    104
United States National Drug Control
    Strategy 91–2

Walker, General Michael 24
Wankel, Doug 95–6
Wikileaks 137
Williams, Michael J. 25, 26, 61
Williams, Zoe 139
Wolfers, Arnold 131
Wood, William "Chemical Bill" 104
Woollas, Phil 29

Yemen 74, 83, 136
Yorke, Claire 152